Alpine Mushrooms
of North America

Alpine Mushrooms
of North America

TREASURES ABOVE TREELINE

Cathy L. Cripps

**UNIVERSITY OF
ILLINOIS PRESS**
Urbana, Chicago, and Springfield

Cataloging data available from the Library of Congress

ISBN 978-0-252-08893-3 (paperback)
ISBN 978-0-252-04837-1 (ebook)

For Don

Contents

Preface

Collecting mushrooms in alpine and Arctic habitats with friends, colleagues, and students has been a serious endeavor for over twenty-five years. This field guide is an attempt to summarize our findings in a format accessible to those interested in the topic. The book covers mushrooms in alpine cirques, basins, passes, and plateaus above treeline in the Rocky Mountains of Montana, Wyoming, and Colorado, with extensions into British Columbia and Alaska. Many of the same fungi are expected in Arctic areas of North America. A significant portion also occur in Arctic and alpine areas of the Alps, Pyrenees, Tatra, and Carpathian Mountains and in Fennoscandia, Greenland, Iceland, Svalbard, the Faroe Islands, the Scottish Highlands, and even a few in Antarctica.

The book is intentionally small so it can be carried into remote areas. Casual observers and naturalists who would like to learn more about the tiny mushrooms found in these places can use it in situ as an identification guide. Alpine wildflower lovers often use a hand lens while lying belly-down on the ground to observe their finds; this works for many alpine "belly" mushrooms as well. The book's detailed descriptions and color photographs can be used to identify the mushroom at hand. For more seriously oriented mycophiles, the book has information on microscopic features, further references, and Genbank numbers.

The book is divided into several ecological groups (mycorrhizal mushrooms, decomposers in meadows and grasslands, and tiny fungi on mosses) to facilitate quick recognition. This format will be unusual to those more familiar with strictly taxonomic layouts. The ecology of the mushrooms is emphasized over rigid family hierarchy.

A primary goal is to bring attention to this overlooked mycoflora for the purpose of conservation efforts. As treeline rises and the Arctic "greens" from climate warming, these precious habitats are being reduced, and a long evolutionary history eradicated. This book contains a summary of what is there or, alternatively, what was once there.

Acknowledgments

Many people made this book possible. First and foremost, I would like to thank my husband, Don Bachman, who was deeply involved in collecting, logistics, and driving for almost every field trip to the Alpine over twenty-five years. I am indebted to Orson K. Miller Jr. for first showing me the alpine mushrooms on Canadian field trips to Banff and Jasper parks with his class at Flathead Lake in western Montana. I thank Meinhard Moser for teaching me the alpine willows and Telemonias one special day on Niwot Ridge, and subsequent interactions. The National Science Foundation funded the initial studies of macrofungi in the Rocky Mountains of Colorado, Wyoming, and Montana. At that time, Egon Horak brought his knowledge of European alpine fungi to North America. Egon, his wife, Almut, Don, and I spent many pleasant days collecting alpine fungi up and down the Rockies.

My graduate students contributed tremendously to this project. Their detailed molecular studies of *Laccaria* (Todd Osmundson), *Lactarius* (Ed Barge), and *Russula* (Chance Noffsinger) made it possible to detect intercontinental distributions. Former graduate student Kate Mohatt led me into the Alaskan outback. Botanist Bill Webber showed me the alpine *Betula* in Colorado. Linnea and Lee Gillman took me to special alpine passes in Colorado. Vera Evenson's high-elevation collections were invaluable. Rick Kerrigan helped with identifying the high-elevation *Agaricus* species, and Pierre-Arthur Moreau molecularly identified the alpine morel. Ellen Larsson and Jukka Vauras were partners in the *Inocybe* work, as were Henry Beker and Ursula Eberhart for the *Hebeloma* study, and Ursula Peintner and Joe Ammirati for *Cortinarius* systematics. Taiga Kasuya's work on *Bovista, Calvatia,* and

Lycoperdon, along with Leo Jalink's notes, are important contributions to the puffballs of the Beartooth Plateau.

I wish to thank my international friends Esteri Ohenoja and Annu Ruotsalainen (Finland), Gro Gulden (Norway), Henning Knudsen and Mikako Sasa (Denmark), Torbjørn Borgen (Greenland, Denmark), Steen Elborne (Denmark), Anna and Michael Ronikier (Poland), Marijka Nauta (the Netherlands), Victor Mukhin and Anton Shirov (Russia), Tomatsu Hoshino and Yuka Yajima (Japan), and Ivano Brunner and Frank Graf (Austria) for exchange of information on Arctic and alpine fungi, mostly during our International Symposia on Arctic and Alpine Mycology (ISAM) meetings. I particularly want to thank Bob and Joanne Antibus for stepping up to help with ISAM 8 on the Beartooth Plateau in Montana; Tim Wheeler for sequencing alpine specimens; and eagle-eyed Else Vellinga for improving the text. This book is offered with deep appreciation of others who have preceded us in the challenging study of Arctic and alpine fungi.

Alpine Mushrooms
of North America

Mushrooms in the Alpine and Arctic Habitats

The alpine and Arctic biome covers a significant portion of the earth's land surface, much of it in the Northern Hemisphere. The true alpine is defined as the open land above treeline on high mountaintops. In North America, treeline reaches 3,200 m (10,500 ft) in the southern Rocky Mountains and descends to almost sea level in the Yukon and Alaska, where the alpine gradually merges with Arctic habitats. The Arctic is defined as the open landscape beyond the trees at high latitudes, or more formally as the area above latitude 66° North within the Arctic Circle. Together the Arctic and the alpine comprise a cold-dominated biome covering 8 percent of the earth's land, with annual temperatures averaging 3.5° F (−16° C) and −1.5° F (−18° C), respectively. Greenland, a part of North America, is not covered specifically in this book because there is much information on its fungi in the European literature. However, many of the species in this book are also found in Greenland.

Why Study Mushrooms in the Alpine and Arctic?

Many of the mushrooms in these cold climates are not found anywhere else on earth, and yet some have wide ranges that include the circumpolar Arctic as well as disjunct mountain ranges across the Northern Hemisphere. For example, the tiny, spoon-shaped *Arrhenia auriscalpium*, only a few millimeters high, occurs on a few isolated mountaintops in the southern Rocky Mountains and is not found at lower or warmer elevations for hundreds of miles around. Yet, this species shows up again in Arctic Greenland, Iceland, Svalbard, Fennoscandia, Canada, Alaska, and the alpine zone of the Alps, Tatras, Pyrenees, and Khibini mountains. Numerous other Arctic-alpine fungi follow this same pattern. Other mushrooms are refugees from the forests below and also occur in boreal and subalpine forests. Molecular methods (DNA analysis) can now confirm that many of the same mushroom species are scattered across all these far-flung cold places. Why? One hypothesis is that historic glaciation pushed Arctic fungi southward, only to leave them stranded on high mountaintops after glacial retreat. Or possibly there was an older alpine mycoflora that stretched across the top of the world and invaded the Arctic when it formed as the climate changed.

Much has been written about these fungi in the Alps, where Arctic-alpine mycology was initiated by Jules Favre in 1955. Subsequently, the mushrooms of Fennoscandia, Greenland, Iceland, and Svalbard were studied in detail. In North America, however, other than a few Arctic fungi picked up by early sailing expeditions, and a few scattered reports of alpine fungi from the Rocky Mountains and Alaska (Noffsinger et al. 2020), little was known of the Arctic and alpine fungi of North America until recently. My research findings of the last twenty-five years, primarily in the Rocky Mountains, are summarized here in accessible field guide format in hopes of inspiring further exploration of these remote areas.

For me, collecting in Arctic-alpine habitats is exciting because these unique cold-loving mushroom species are rare yet widely distributed. Some of the mushrooms my crew has discovered are "first reports" for North America, and yet many are also known from around the Arctic Circle and from other mountain ranges in the Northern Hemisphere. Some are new species. Walking across tundra on a warm sunny day, moving through a dreamscape of alpine flowers, and plucking tiny fungal gems nestled in moss, grass, and willows is a thrilling and scientifically satisfying pastime.

The alpine and Arctic are not places to fill your basket with edibles. These precious species should be considered for conservation, especially since climate change is already warming these habitats, alpine treeline is rising, and the Arctic is "greening" with shrub encroachment.

What Are the Challenges?

In the alpine, the field season is short, lasting only a few weeks from about the third week of July to the end of August. Alpine air is thin and holds less oxygen for every foot or meter gained in elevation; stamina is required to reach these heights. UV light is strong at altitude, and protection in some form is a necessity, whether it be sunscreen or a hat and appropriate clothing. Alpine summer days can be almost hot and then dip to below freezing within a short period of time; nights are almost always chilly. Carrying a coat is necessary. Frost and snow are possible throughout the summer. Snow can stimulate fruiting, but it also buries the mushrooms. Summer thunderstorms are dangerous, as there are no places to hide from lightning strikes in the alpine. It is best to collect in the morning and avoid afternoon thunderstorms. The alpine can be exceptionally dry at times due to lack of rain and desiccating winds—this is called a "brown out." It is not worth collecting during these dry periods. Forest fires at lower elevations can add smoke to the pristine alpine air and block access to higher elevations during fire season. Insects, especially mosquitoes and biting flies, are especially numerous in the early part of the alpine field season.

The Arctic has its own challenges. In the land of the midnight sun, diurnal temperature fluctuations are reduced, tempering the climate. But constant summer sunlight can be disorienting time-wise, and you might find yourself collecting well into the evening hours. A mask can be used at night to regulate sleep. Fog and mist can quickly obscure the landscape and require navigating blind or hunkering down in place—so be prepared. Muskeg and quicksand should be avoided, and crossing cold rivers and wet landscapes should be done with caution. In some Arctic areas, an awareness of muskoxen, polar bears, and grizzly bears is crucial; polar bear guards are required in Svalbard, and elsewhere it may

be necessary to bring bear spray or bear deterrents for brown bears or grizzlies. Mosquitoes and biting flies can occur in overwhelming numbers throughout the field season in the Arctic, and face nets are then essential.

The bottom line is: always be prepared when venturing into these harsh habitats, and make sure to obtain a permit where appropriate.

When Is the Best Time to Look?

The field season is short, and its initiation varies with winter and early spring weather conditions. Typically, a few mushrooms will start fruiting in late July (often the third week). This includes species of *Arrhenia*, *Galerina*, and *Entoloma*. However, deep snow and low temperatures can delay fruiting. The soil needs to warm, and lakes and snowbanks need time to melt after the winter. This occurs more quickly in the Arctic with increasing daylight; diurnal temperature fluctuations in the alpine mean cold nights, which slow the warming process. *August is the month to collect Arctic and alpine mushrooms!* And the peak time will depend on prior and present weather conditions. In early September, the willows and birch turn color and drop their leaves as the season comes to a quick end.

Where Are the Best Places to Look?

Mushrooms can be almost anywhere in these cold environments, but they are often concentrated in certain localized habitats, some of which are the major divisions in this book: willow and *Dryas* mats, clumps of willow shrubs, birch copses, meadows, and mossy banks near streams and ponds.

Mycorrhizal (ectomycorrhizal) mushrooms are associated with particular host plants, so concentrating collecting efforts around willows (*Salix*), birch (*Betula*), and *Dryas* is the best way to find them. The taller shrub willows (and birch and alder) are easy to spot across the landscape, as they are the tallest vegetation around. Mycorrhizal mushrooms can be found on the edges or interior of willow clumps. Small decomposers sometimes fruit on bare soil or debris deep inside willow clumps. This necessitates either crawling or wading into the center of the taller willows. We dubbed this search technique "willow diving."

Dryas and dwarf willows produce low, sometimes extensive mats. The best way to locate them is to learn to recognize subtle color patterns across the landscape. Variations and patchiness can be discerned from a distance with binoculars. Sometimes these mat plants are near or mixed with taller shrub willows in wetter areas. Some mycorrhizal mushrooms in these mats are so small that it is best to lie down at the edge of the mat to view and collect them. Hence the name "belly mushrooms."

In open grasslands and meadows, the decomposer mushrooms and puffballs are more dispersed. Many tend to be somewhat large, some species form rings, and many are white, all of which makes them easily visible when walking through open areas. Mosses concentrated around seeps, streams, and ponds are good places to search for the

Willow clumps and mats across the landscape.

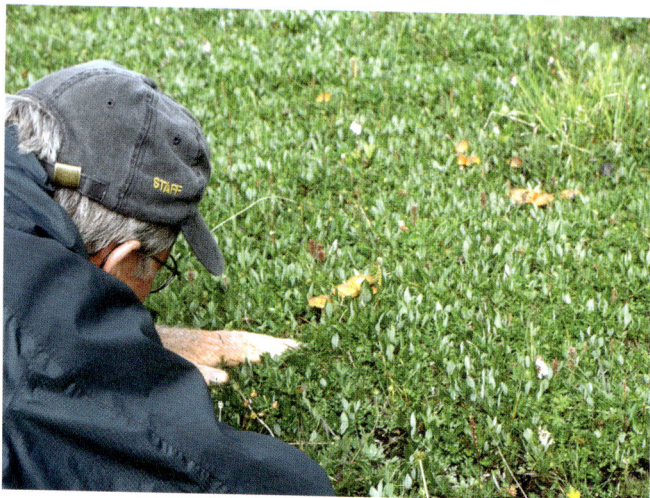

Viewing "belly mushrooms" in the Alpine.

Collecting below a melting snowbank.

miniature fungi associated with them. Other places to look include protected microhabitats, such as rocky ledges and sunken areas out of the wind, and north-facing slopes. Remnant snowbanks provide meltwater for the vegetation below. This can be a good place to look for Arrhenias on bare soil or mycorrhizal mushrooms near willows. If there are minor differences in elevation on a slope, it can be useful to find the "fruiting zone." If fungi are fruiting at a certain level (and not above or below), more fungi can be discovered by moving along this zone horizontally.

How Are Arctic-alpine (AA) Mushrooms Examined for Study?

Casual observers may only wish to view the alpine mushrooms in the same way they view alpine wildflowers. In this case, a photo may be sufficient as a record and for identification. Note the general habitat. Are the mushrooms clearly in the alpine zone or near treeline? Are they clustered or scattered in grass, moss, *Dryas*, or with willow or birch? Note the location, elevation, and date observed.

Often a sample is needed for further examination and/or microscopic work. This requires carefully removing a few mushrooms of various ages, making sure to retain the base of the stem (and in some cases, without touching the stem!). Most mushrooms in the alpine and Arctic are small, and it is best to use something like a fishing tackle box with divisions for collecting the fragile fruiting bodies. Small bits of moss or lichen can be placed in the bottom of each section to keep the specimens moist. Mushrooms dry out quickly in wind or extreme temperatures. Specimens should be photographed in their natural habitat if possible, although cold, wind, or insects can make field photography impossible. Only pristine mushrooms should be collected, as wind and frost can obscure features important for identification. If possible, transport collections in a cooler on longer drives. Keep conservation ethics in mind if collecting actual specimens.

Each collection should be described while fresh (see next section), before delicate features and odors fade. Additional photographs can be taken on gray cards. Then specimens should be placed on a dehydrator with a label as quickly as possible, dried overnight, and packaged in the morning. Be sure to consider which drying methods are possible for

Collecting boxes for small alpine mushrooms.

the situation. Dehydrators are ideal but electricity is required. Screens, gas lanterns, and silica gel can suffice for the smaller specimens. When dry, specimens are packaged with labels for transport. Newspaper or small paper or plastic bags can be used as packaging.

How Are the AA Mushrooms Described?

Fungal organisms are comprised of tiny threads of hyphae (mycelium taken as a whole) usually hidden in the soil or substrate. The fungus derives its nutrition from mycelial processes, whether it is decomposition of dead matter or acquiring nutrients from a living host. The fruiting body (i.e. mushroom) is a reproductive structure designed to lift the fungus out of the substrate and produce spores to spread the organism. The features of the fruiting body are important for identification, but those of mycelium are not.

Mushrooms can be described in the field, but more typically this is done upon return to base camp, lab, or home. The size, shape, color, and surface texture of the cap and stem should be recorded, as should the color and sometimes the number of gills. If possible, determine the odor in the field while the mushrooms are fresh. Odor can be particularly important for identifying certain species. Once the mushroom has dried, this feature will be obliterated. Notice how the descriptions and photographs in this book are used together to represent the fresh mushrooms. Pay attention to noting particular characteristics for each genus, such as milk color and gill staining for Lactarii, taste for Russulas, and odor for Inocybes. Again, don't forget to include field notes as to habitat, location, and date.

It is important to know spore color to use some of the keys in this book; this is often difficult to determine with small alpine mushrooms. A spore print may be necessary to determine whether the mushroom has light or dark spores, or the exact

Morphological features of a mushroom.

color. This can be done efficiently in the field by placing a cap on a small piece of white paper, wrapping it in tinfoil, and placing it gill-side down in your box. Ideally, the mushroom will spore print while still in the field or in transit. Alternatively, spore color can sometimes be observed on the stem, or on overlapping caps. Spore color is not always the same as gill color.

A microscope can help determine or confirm identifications. Magnification of a small piece or cross-section of gill can reveal spores, basidia, and cystidia. Basidia are cells that hold the spores (usually four), and cystidia are distinctively shaped sterile cells on the gill faces and edges. Scalps of stems and caps can reveal cystidia on these surfaces as well. A microscope equipped with an oil immersion lens (1000X) and a micrometer are useful for measuring spore size. A 3 percent solution of KOH and a vial of Melzer's reagent are helpful.

A glossary that includes macroscopic and microscopic features is located at the back of the book.

Spore printing is used to determine spore color for identification.

Evaluating mushroom odor in the field.

Collecting via helicopter in Alaska.

How to Use This Book

There are several ways to use this book, depending on whether you are a beginner or are more experienced at collecting fungi in cold habitats.

To identify a mushroom, *beginners, naturalists,* and *casual observers* first can see if it fits into one of the five main sections of this book and go to that page to start the process.

Alternatively, you can start with the **Field Key** to mushroom genera below. For this process, a spore print may be necessary, because you will need to know whether the mushroom has light or dark spores (see previous section). Then begin the identification process by selecting each appropriate alternative in a couplet.

More experienced collectors may find that using the **list on the inside cover** is the best way to find a specific genus. The **index** covers all the individual species. For more in-depth information, refer to the **references** for each species, which give a more detailed description and microscopic data. Mushroom names follow **Index Fungorum** (Index Fungorum Home Page).

The **collection number** at the bottom of each page (i.e. CLC 001) refers to the mushroom in the photo. CLC (Cathy L. Cripps) collections are currently deposited in MONT Herbarium (Montana State University); DBG specimens at the Denver Botanic Garden; and ZT collections at Herbarium der Eidgenössische Technische Hochschule Zürich, Switzerland. The ITS region of the DNA has been sequenced for some collections, and results deposited in the Genbank database (GenBank Overview at nih.gov) for reference. The **Genbank number** is given after the collection number at the end of selected mushroom descriptions. Sequences not currently available are labeled simply as "sequenced."

Key to Gilled Alpine Mushrooms

The main sections of the book contain keys to species in each habitat, which is another way to start the identification process. A key to non-gilled fungi is on page 17. For determining spore color, see page 9.

WITH VOLVA OR A RING ON THE STEM

1. Base of stem with a cup (volva) or rings of tissue
   Amanita (pg 87)
1. Not as above 2

2. Stem with a membranous ring (can be minimal) 3
2. Ring absent, or with a fibrous/cobwebby ring 6

3. Gills free (not touching stem) 4
3. Gills attached to the stem 5

4. Cap white to buff; gills pink to brown; spores brown
   Agaricus (pg 26)
4. Cap white to pinkish, scaly or smooth; gills and spores
   white **Lepiota** (pg 37)

5. Cap orange, granular; stem scaly; spores white **Cystodermella** (pg 33)

5. Cap buff, yellow brown, smooth; spores brown **Agrocybe** (pg 28) (also see *Galerina pseudomycenopsis, Stropharia alpina*)

SMALL TYPES; CAPS BELL-SHAPED OR CONVEX

6. Small types; cap bell-shaped or convex, often striate; stem thin ≤ 2 mm wide; in moss or *Dryas* 7

6. Cap funnel- or fan-shaped *or* stem wider; can be in moss or *Dryas* 12

7. Spores brown, purple-brown, or purple-black; gills orange or blackish 8

7. Spores white; gills white, yellow, buff 9

8. Cap orange, orange-brown; gills orange; in moss; spores brown **Galerina** (pg 68)

8. Cap red-brown; gills blackish; spores purple-black; in moss **Deconica** (pg 67)

9. Cap bell-shaped, striate; stem long, thin, fragile or not; in moss 10

9. Cap convex, pleated or not; stem black and tough or fleshy; in *Dryas* 11

10. Cap olive or yellow, striate; stem fragile **Mycena citrinomarginata** (pg 77)

10. Cap gray, striate; stem somewhat tough **Mycena cf. pasvikensis** (pg 78) (if cap lavender, see *Mycena pura* nanoform, pg 45)

11. Cap pleated; stem black, tough **Rhizomarasmius epidryas** (pg 80)

11. Cap smooth; stem pale buff, fuzzy at base **Gymnopus alpicola** (pg 34)

SMALL TYPES; CAPS FUNNEL-SHAPED OR FAN-SHAPED

12. Small types; cap funnel- or fan-shaped; gills decurrent; stem ≤ 2 mm wide; in moss, on liverworts, or on soil 13
12. Stem wider; gills decurrent or not 18

13. Cap, gills and stem bright yellow; on soil **Lichenomphalia** (pg 40)
13. Cap another color 14

14. Cap orange, deep orange; gills white to orange 15
14. Cap another color 16

15. Stem short, orange; on liverworts **Loreleia** (pg 75)
15. Stem long, thin, pale, rubbery; in moss **Rickenella fibula** (pg 81) (also see *Laccaria*)

16. Gills as veins; stem absent or very short **Arrhenia** (pg 63)
16. Gills normal; stem present, can be short 17

17. Cap brown, striate; stem pale, long; in moss **Omphalina rivulicola** (pg 79)
17. Cap gray, blackish brown, striate; stem short; on soil or moss **Arrhenia** (pg 65) (if cap white, see *Clitocybe dryadicola*)

FLESHY TYPES; CAPS CONVEX OR DEPRESSED; SPORES WHITE, YELLOW, PINK

18. Spores white, yellow, pink 19
18. Spores brown, yellow-brown, rusty, blackish 28

19. Cap mostly gray, gray-brown, black-brown; gills pink to white; spores pink, angular; usually near willows **Entoloma** (pg 118)
19. Not as above; spores white to yellow, not angular 20

20. Cap small, dark to pale orange; gills pink to pale orange; spores white, spiny; near willows **Laccaria** (pg 178) (also see *Lactarius lanceolatus*)
20. Not as above 21

21. Cap bright yellow, orange-red, waxy, viscid; spores white, smooth; in grass **Hygrocybe** (pg 35) (also see *Lactarius salicis-reticulatae, Cortinarius vibratilis*)

21. Not as above 22

22. Cap red, orange-red, magenta, maroon, violet, lavender, purple 23

22. Cap white, yellow, golden, buff, grayish brown, orange-brown, pale orange 25

23. Cap vinaceous purple, velvety; gills yellow **Calocybe onychina** (pg 30)

23. Not with all the above features 24

24. Cap, gills, and stem lavender; spores white **Mycena pura** group (pg 45) (some *Cortinarius* species have lavender colors but brown spores)

24. Cap red, orange-red, magenta, maroon, violet; gills white, gold; stem white to pink, brittle; spores white/yellow with amyloid warts; willow **Russula** (pg 198)

25. Gills giving milk or staining when cut; cap slightly sunken in center; gills slightly decurrent; spores white, warts amyloid; near willows **Lactarius** (pg 186) (if cap is white, also see *Russula laevis*)

25. Not as above; usually in grass or *Dryas* 26

26. Cap large, rather flat with umbo, cream, pale orange; gills attached; stem slim, tall, twisted striate; spores with amyloid warts **Melanoleuca** (pg 41)

26. Not as above; gills often more decurrent 27

27. Cap thin, orange-brown or white, depressed; gills decurrent **Clitocybe** (pg 31)

27. Cap and stem fleshy, white to buff; gills slightly decurrent or not **Lepista** (pg 39) (if cap white with brown spores, see next section)

FLESHY TYPES; SPORES BROWN, YELLOW-BROWN, RUSTY, BLACKISH

28. Cap turquoise, buff, or olive-yellow; spores purple-black 29

28. Not as above; spores yellow-brown, rusty, brown; often with cortina 30

29. Cap turquoise to buff; gills dark; can have slight ring; in grass **Stropharia aplina** (pg 46)

29. Cap, gills, and stem olive-yellow, smooth, greasy; in moss **Hypholoma** (pg 74) (also see *Deconica*)

30. Cap brown, scaly; stem base blue; odor fishy **Inosperma calamistratum** (pg 170)

30. Not as above 31

31. Cap or stem typically rough, scaly, ocher, golden, red-brown, or covered with white tissue; odor often of burnt sugar **Mallocybe** (pg 168)

31. Not as above 32

32. Cap radially fibrous to smooth, usually conic-convex; odor spermatic, a few fragrant-pungent; spores smooth or nodulose **Inocybe** (pg 142), **Pseudosperma** (pg 168)

32. Not as above; cap usually smooth, convex, flat, a few conic pointed 33

33. Cap whitish to brown, some 2-toned, smooth, greasy, convex or flat; odor often of radish; spores brown, smooth or slightly bumpy **Hebeloma** (pg 128)

33. Cap white, buff, light/dark brown, yellow, orange, some lavender tinted, usually smooth, viscid to dry; spores brown, red-brown, warty **Cortinarius** (pg 92) (also see *Agrocybe praemagna*)

Key to Non-Gilled Alpine Fungi

1. With cap and stem 2
1. Not as above 3

2. Cap smooth; pores/tubes under cap; stem rough
   Leccinum (pg 196)
2. Cap pitted and ridged; stem smooth or bumpy **Morchella**
   (pg 43)

3. Fruiting body club or disc-shaped 4
3. Not as above 7

4. Fruiting body disc-shaped 5
4. Fruiting body more or less club-shaped 6

5. Stem present; disc brownish **Ciborinia** (pg 216)
5. Stem absent; disc orange-red **Scutellinia** (pg 217)

6. Tiny, orange-brown, with small head and stem, in moss
   Bryoglossum (pg 214)
6. Larger, yellow, club-shaped, in clusters in heath **Clavaria**
   (pg 218)

exoperidium

gleba with spores

base

Structure of a puffball.

7. On the ground, in grass; round, oval, pear-shaped
   Bovista, Lycoperdon (pg 51)
7. Not as above; on wood or leaves 8

8. Orange pustules on willow leaves **Melampsora** (pg 220)
8. Small bird's nest-type structures on willow wood
   Crucibulum (pg 219)

Gilled Mushrooms by Cap Color

White (can be buff in age)
Agaricus
Amanita arctica/nivalis
Clitocybe dryadicola
Collybia irina
Cortinarius albidipes
Hebeloma
Inocybe pallidocremea
Inocybe pudica/geophylla
Inocybe fraudans
Lepiota erminea
Lepista irina
Leucocalocybe
Melanoleuca
Pseudosperma bulbosissimum
Russula laevis

Yellow
Cortinarius vibratilis
Hygrocybe
Lactarius repraesentaneus
Lactaricus salicis-reticulatae
Lichenomphalia alpina
Mycena citrinomarginata

Yellow-brown, ocher, olive
Agrocybe
Arrhenia

Cortinarius absarokensis/alpinus
Cortinarius cinnamomeoluteus
Hypholoma
Inocybe
Lactarius repraesentaneus
Mycena citrinomarginata
Pseudosperma cf. *flavellum*
Mallocybe

Orange
Cortinarius alpinus
Cortinarius parvannulatus
Cystodermella
Galerina
Hygrocybe
Laccaria
Lactarius lanceolatus
Loreleia
Melanoleuca
Rickenella
Russula intermedia

Orange-brown
Clitocybe festiva
Cort. absarokensis/alpinus
Cortinarius cinnamomeoluteus
Cortinarius ferruginosus
Cortinarius pratensis

Cortinarius uliginosus
Mallocybe
Russula subrubens/intermedia

Red, pinkish
Hygrocybe
Lepiota favrei
Russula nana

Lavender, magenta, purple
Calocybe onychina
Cortinarius
Lactarius brunneoviolaceus
Mycena pura
Russula

Turquoise fading buff
Stropharia alpina

Buff, light brown, brown
Agrocybe
Amanita groenlandica
Arrhenia
Cortinarius
Deconica
Gymnopus alpicola

Hebeloma
Inocybe
Lactarius glyciosmus
Lactarius pallidomarginatus
Lactarius nanus
Melanoleuca
Omphalina rivulicola
Pseudosperma
Rhizomarasmius

Dark brown
Arrhenia velutipes
Inocybe
Inosperma calamistratum
Cortinarius hinnuleus
Cortinarius saturninus
Cortinarius piceidisjungendus

Gray, blackish gray
Amanita nivalis
Arrhenia
Mycena pasvikensis
Entoloma
Hebeloma subconcolor

1

Mushrooms of Meadows and Grasslands

Meadows and grasslands above treeline are often considered the very definition of alpine, with open wind-swept vistas and an abundance of alpine wildflowers, such as the deep blue alpine forget-me-nots, bright pink moss campion, white arctic gentians, and yellow alpine avens. Although sparsely distributed, meadow mushrooms are some of the largest and most colorful mushrooms above treeline. They are often discovered serendipitously when mushroom seekers hopscotch between wet willow beds and mossy seeps where mushrooms are concentrated. Open grassy vegetation beneath melting snowbanks is a good place to look for these mushrooms.

Alpine meadows and grasslands support fairy rings of *Agaricus, Melanoleuca, Lepista,* and puffballs; rings can be large because growth is unimpeded in the vast landscape. Many typical grassland mushrooms transcend treeline because they are not tied to living woody vegetation but are decomposers. Mushrooms here can be as colorful as the alpine wildflowers. Hygrocybes, indicators of pristine grasslands, come in bright red, yellow, and orange but are rather rare in the semi-xeric Rocky Mountains. An unusual turquoise *Stropharia* and a pink *Lepiota* live here, as do lavender Mycenas. Nestled in the occasional patch of *Dryas* are species of *Cystodermella, Clitocybe,* and *Gymnopus*. There is even a rare, high elevation, cold-loving morel. Mushrooms may sprout from the dung of herbivores, but these are not described here as many are brought in by domestic animals from lower elevations.

Decomposers

Mushrooms of Alpine Meadows and Grasslands (some in *Dryas*)

1. Cap with pits and ridges; stem hollow; gills absent **Morchella**
1. Cap, stem, and gills present 2

2. Cap rather pale: white, cream, buff, orange-cream, yellow-brown in one 3
2. Cap more highly colored 12

CAP WHITE, BUFF, ORANGE-CREAM, YELLOW, YELLOW-BROWN, IN GRASS

3. Spores dark-colored; stem usually with a ring 4
3. Spores light-colored; stem with ring or not 7

4. Gills free, pink, grayish, dark brown; ring membranous 5
4. Gills attached, buff, brown, orange-gray; ring fibrous or absent 6

5. Cap white to buff, fibrillose; veil copious **Agaricus cf. altipes**

5. Cap white, smooth or with rough scales; veil as a band **Agaricus chionodermus**

6. Cap yellow-brown; gills red-brown; stem with ring **Agrocybe praecox**
6. Cap white, buff; gills orange-gray; stem without ring **Agrocybe praemagna**

7. Gills free; cap slightly fibrillose or scaly; ring usually present 8
7. Gills attached or decurrent; cap smooth; ring absent 9

8. Cap white, fibrillose; stem scaly with ring zone **Lepiota erminea**
8. Cap whitish, pinkish brown, scaly; ring membranous **Lepiota favrei**

9. Gills decurrent; cap and stem white; in *Dryas* **Clitocybe dryadicola**
9. Gills attached; in grass 10

10. Cap white, margin rolled-in; stem fleshy **Lepista irina var. montana**
10. Cap white, orange-cream; stem thin, striate-twisted 11

11. Cap pale orange-cream; lots of pointed cystidia present **Melanoleuca cognata**
11. Cap grayish to white; cystidia scarce **Melanoleuca excissa**
    (If cap white, also see certain *Hebeloma* and *Inocybe* species.)

CAP ORANGE, ORANGE-BROWN, IN GRASS OR DRYAS

12. Cap orange, orange-brown 13
12. Cap another color 15

13. Cap smooth, orange-brown; gills decurrent; ring absent **Clitocybe festivoides**
13. Cap scaly or granular; with ring or ring zone 14

14. Cap scaly, pinkish brown when young; gills free **Lepiota favrei**
14. Cap granular, orange-brown; gills attached **Cystodermella adnatifolia**

CAP YELLOW, RED, REDDISH ORANGE, SMOOTH, GREASY, IN GRASS

15. Cap yellow, red, reddish orange, greasy 16
15. Cap turquoise, burgundy, lavender, purplish 17

16. Cap convex, orange-red; stem robust, fleshy **Hygrocybe punicea**
16. Cap tall, pointed, yellow, turning black; stem thin **Hygrocybe conica** (if yellow, also see *Lichenomphalia alpina*)

CAP TURQUOISE, BURGUNDY, LAVENDER, PURPLE

17. Cap and stem bluish green to buff; spores purple-black **Stropharia alpina**
17. Cap burgundy, lavender, purplish; spores white 18

18. Cap burgundy; gills yellow; stem wine-color **Calocybe onychina**
18. Cap lavender to purple; gills lavender **Mycena pura, Mycena diosma** (some *Cortinarius* species have lavender colors but brown spores)

Mushrooms That Can Occur with *Dryas*

DECOMPOSERS

Calocybe onychina
Clitocybe dryadicola
Clitocybe festivoides
Gymnopus alpinicola

Lepiota favrei
Morchella norvegiensis
Mycena pura
Rhizomarasmius epidryas

Mycorrhizal mushrooms can also occur in *Dryas*.

Alpine Meadow.

Agaricus cf. *altipes* (Moeller) Pilát

Cap 4.5–7 cm wide, hemispherical, slightly squarish, then convex, broadly convex to almost flat with dome, white to cream, later pale buff, smooth in center, outward appressed fibrillose, with hairs at margin; margin hung with floccose veil tissue, which can be copious. **Gills** free, narrow, bright pink at first, then grayish pink, finally dark brown. **Ring** superior, membranous, thick cottony, white, tissue mostly remaining on cap margin. **Stem** 4–6 × 0.8–2 cm, rather long, equal, slightly pointed at base, white, staining a bit brown, with raggedy floccose zones below ring. **Flesh** white, staining slightly brown. **Odor** not distinct, possibly fungoid. **Spores** brown, 8–10 × 4.5–6 µm, ellipsoid, smooth, no germ pore.

Ecology and Distribution In open grassy meadows at 3,810 m in the San Juan Mountains of Colorado. Also known from high-elevation conifer forests in New Mexico, and lower elevations in Washington and Alaska. July-August.

Notes The fibrillose cap and copious veil are diagnostic features. *A. altipes* is a European species that fits our Rocky Mountain collections, but there is no sequence data for the European type collection. The sequence of our collection was examined by R. Kerrigan.

- Cap white turning buff, appressed fibrillose
- Gills free, pink, then brown
- Ring white, membranous, copious
- Stem white floccose below
- Spores brown
- In open grassy meadows

Reference Kerrigan 2016. CLC 1662, Genbank PQ165818.

Agaricus chionodermus Pilát

Cap 4.5–9 cm wide, fleshy, convex, broadly convex, white, buff in center, smooth to slightly scaly or developing flat, coarse scales; cuticle extends beyond gills. **Gills** free, pink-gray, brown in age. **Ring** membranous, flaring up or down, floccose but soon as a small band. **Stem** 3–6 × 1.0–2.5 cm, robust, white, equal, pointed at base, smooth above ring, minutely floccose below. **Odor** astringent to fungoid. **Flesh** firm, white, no color change, staining a bit brown. **Spores** brown, 8–10(–12) × 5–6 μm, ellipsoid, smooth, no germ pore.

Ecology and Distribution Common on the Beartooth Plateau in Montana and Wyoming above and below treeline, in open meadows. Also known from New Mexico, Colorado, British Columbia, and Alaska, primarily in conifer forests. July-August.

Notes This species, typically of conifer forests, apparently reaches into low alpine meadows. It has likely been mistaken for a robust, scaly *A. campestris*. The sequence was examined by R. Kerrigan.

- Cap white, scaly at first, scales becoming flat
- Gills free, pink then dark brown
- Ring floccose and then as a band
- Stem floccose below ring
- Spores brown
- In open meadows

Reference Kerrigan 2016. Above CLC 1982b. CLC 1983b, Genbank PQ156405.

Agrocybe praecox (Pers.) Fayod complex

Cap 2–5 cm across, convex with a dome or low umbo, can become sunken in center, pale ocher, yellow-brown, smooth, greasy when young, kidskin dry in age; margin with veil tissue. **Ring** membranous, superior, flaring up or down, dingy white, buff. **Gills** emarginate, pale gray-brown, brown, red-brown. **Stem** 3–5 × 0.4–0.8 cm, equal or larger toward base, dingy whitish, buff, striate. **Flesh** whitish, firm; stem stuffed. **Odor** farinaceous. **Spores** brown, 9–11 × 5–7 µm, ellipsoid, smooth, with large germ pore; cystidia narrow at top and swollen at base, scarce.

Ecology and Distribution In the alpine, occurring in open grasslands on soil; reported from Independence Pass (3660 m), Colorado, the Beartooth Plateau (3200 m), Montana, and the Austrian Alps. A widely distributed complex better known from lower non-AA habitats. August in the alpine.

Notes *Agrocybe* species are not common in alpine situations, but the *A. praecox* complex has also been reported near treeline in Austria and on the Beartooth Plateau also by M. Moser. *A. praemagna* is larger and lacks a ring.

- Cap buff to yellow-brown, kidskin smooth
- Gills attached, brown
- Stem whitish buff with membranous ring
- Spores brown with germ pore
- In open grasslands and alpine meadows

References Moser and Horak 2006; Moser 1986. CLC 1576.

Agrocybe praemagna M.M. Moser & E. Horak

Cap 5–10(–15) cm wide, broadly convex, shallow convex, dingy white to pale brown, leather color, sometimes with pale brown spots, smooth, suede-like in age; margin not striate. **Ring** usually lacking, even in young specimens. **Gills** emarginate to decurrent, crowded, pale then darker orange-gray, gray-brown. **Stem** 5–10(–14) × 1–1.5(–2) cm, enlarged toward base, dingy whitish ocher, browning with age, rough-fibrous, dry. **Flesh** whitish, solid. **Odor** indistinct or faintly farinaceous. **Spores** blackish brown, large, 13–18 × 7–10 µm, ellipsoid, with germ pore; cystidia variable.

Ecology and Distribution On soil in open alpine meadows just above or below treeline. Known from the Teton Mountains of Wyoming, Independence Pass in Colorado, and the Beartooth Plateau of Montana; also in aspen forests. Possibly indigenous. August.

Notes *Agrocybe* species are recognized by attached gills, brown spores with a germ pore and a cracking cap due to a cellular cuticle. *Agrocybe* species typically have a membranous ring that is absent in this large species.

- Cap robust, whitish to pale brown, smooth
- Gills orange-brown to brown, attached
- Stem robust, whitish or slightly yellow-brown, striate
- Ring absent
- Spores blackish brown
- In meadows just above or below treeline

Reference Moser and Horak 2006.

Calocybe onychina (Fr.) Donk

Cap 3.5–4.5 cm wide, broadly convex, dull vinaceous, burgundy, lavender, minutely velvety. **Gills** attached, crowded, pale to dark golden yellow. **Stem** 2–4 × 2 cm, whitish with vinaceous tones, equal, fibrous, fleshy. **Flesh** yellowish white. **Odor** indistinct or faintly farinaceous. **Spores** white, tiny, 4–4.5 × 2–3 μm, ellipsoid, smooth.

Ecology and Distribution Here reported from the alpine (Loveland Pass, Colorado, 3,620 m) nestled in dwarf willows and *Dryas*. Typically found in high elevation conifer forests of the West and also known in Europe. Somewhat rare. August in the alpine.

Notes Also known as *Rugosomyces onychinus*. A striking mushroom with purple velvety cap and golden yellow gills. Not usually reported from the alpine. A rare fungus.

- Cap burgundy, velvety
- Gills golden yellow
- Stem pale wine-color
- Spores white
- In *Dryas*

Reference Desjardin, Wood, Stevens 2015. ZT 8073.

Clitocybe dryadicola (J. Favre) Harmaja

Cap 3–4 cm wide, shallowly dished, concave, whitish with hoary coating, cracking somewhat in center; margin thin. **Gills** slightly decurrent, white to cream, somewhat crowded. **Stem** 3.5–4.5 × 0.5 cm, rather long, slightly larger at base, white, fibrous, floccose at apex. **Flesh** cream, thin in cap. **Odor** faint. **Spores** 4.5–5.5 × 3–4 µm, white, ellipsoid, smooth.

Ecology and Distribution Reported in the Rocky Mountain alpine near Banff National Park and on Loveland Pass, CO, in *Dryas*. Also known from the Alps above treeline with *Dryas*. August.

Notes The association with *Dryas* is assumed not to be mycorrhizal since Clitocybes are decomposers. The thin, white frosty cap, decurrent gills and long, slender stem are diagnostic. The similar *C. candicans* is typically on litter or wood at lower elevations. *C. lamoureae* has rose tints and narrower spores.

- Cap white, frosty, slightly dished
- Gills white, decurrent
- Stem slender, long, white
- Spores white
- In *Dryas*

Reference Favre 1955. CLC 2299.

Clitocybe festivoides Lamoure

Cap 2–5(–9) cm wide, hemispherical, convex, broadly convex, funnel-shaped in age, light brown, light orange-brown, drying buff, smooth, greasy, fatty-shiny; margin turned in at first, then uplifted, often with a white rim. **Gills** decurrent, cream color, crowded. **Stem** 2–4 × 0.4–0.6 cm, equal or narrowing at base, watery orange with whitish covering. **Flesh** cream. **Odor** fragrant. **Spores** 5 × 3 µm, smooth, ellipsoid, inamyloid.

Ecology and Distribution Densely cespitose in grass, often with *Dryas* in alpine habitats, sometimes forming arcs; from the Beartooth Plateau in Montana, and other Arctic-alpine habitats in Iceland, Norway, Svalbard. August.

Notes Similar to *C. festiva,* which is also alpine, but has a gray-brown cap and lacks an odor. High-elevation *Gymnopus* species can have smooth, orange-brown caps, but stems are thinner and more brittle. Our specimens may have been too fresh to have the reported disagreeable odor.

- Cap convex, orange-brown, smooth, greasy
- Gills whitish, crowded
- Stem pale orange
- Spores white
- In dense clusters in open meadows and grasslands

References Armada et al. 2024. Lamoure 1972. CLC 1990.

Cystodermella adnatifolia (Peck) Harmaja

Cap 2–4.5 cm wide, broadly convex, to almost flat or even sunken in center, orange, orange-brown, with fine granules in concentric circles, background white; margin with white veil remnants. **Gills** attached, crowded, white, cream. **Ring** as floccules, defining a ring zone. **Stem** 1.5–4(–6) × 0.5–1.0 cm, enlarging toward base, pale orange, smooth above ring and with orange floccules below; white at base. **Flesh** white, cream. **Odor** fruity, fragrant. **Spores** white, small, 4–5 × 3 µm, ellipsoid, smooth, not amyloid; no cystidia.

Ecology and Distribution Occurring on Independence Pass, CO, in alpine vegetation including willows and birch. Known mostly from subalpine and boreal forests in North America and elsewhere but also reported from Arctic-alpine habitats in Greenland, Canada, Alaska, and Europe. August.

Notes *Cystoderma arcticum* and *Cystodermella granulosa* var. *granulosa* have also been reported from AA habitats; the former is smaller and paler, and the latter is similar to our species with narrower spores. *C. granulosa* dries more brown, *C. adnatifolia* more orange.

- Cap orange-brown, with fine granules
- Gills white, attached
- Stem with orange floccules below ring zone
- Spores white
- In alpine vegetation

References Smith and Singer 1945; Saar et al. 2009. CLC 3785.

Gymnopus* cf. *alpicola (Bon & Ballará) Esteve-Rav. et al.

Cap 0.5–1.5 cm wide, convex with rolled-under margin, light brown, buff, pinkish buff, darker where bruised, smooth, dry, very weakly striate. **Gills** adnexed, cream to clay color, slightly separated, thickish; edges floccose. **Stem** 2–3.5 × 0.1–0.2 cm, long and thin, equal or slightly larger toward base, flattened, pale orange at apex, below white tomentose to base over darker colors; attached to *Dryas*. **Flesh** pale orange, stem tough, rubbery. **Odor** of garlic or rotten cabbage. **Spores** white, 6–8 × 3–4 µm, ellipsoid smooth, inamyloid.

Ecology and Distribution In dense clusters in *Dryas* in open areas on the Beartooth Plateau (3200 m). Known from alpine areas of the Pyrenees and the French Alps with *Dryas*. Rare. August.

Notes This small *Gymnopus* is recognized by the white tomentose stem, odor of rotten cabbage or garlic, and alpine habit. No bluish was noted in the stem as in European collections, but the stem was bluish under an Ott light. Molecularly close to *G. barbipes*.

- Cap small, buff pinkish, smooth
- Gills cream to clay color
- Stem long and thin, covered with white tomentum
- Odor of garlic or rotten cabbage
- Spores white
- In *Dryas*, in clusters

References Esteve-Raventós et al. 1998. Knudsen and Vesterholt 2008. CLC 3522, sequenced.

Hygrocybe conica (Schaeff.) P. Kumm. complex

Cap 1.5–2 cm wide, strongly conic-convex with tall, pointed umbo, bright yellow with golden orange tones, turning black from the umbo downward, smooth, greasy. **Gills** almost free, white then yellow, staining black. **Stem** 5–6 × 0.3–0.4 cm, long, bright luminescent yellow, turning black from the base up, striate. **Flesh** yellow, blackening. **Odor** none. **Spores** white, 9–11(–13) × 5.5–6.5 µm, smooth, ellipsoid.

Ecology and Distribution Reported in the alpine and Arctic, in grasslands and in alpine vegetation. Known from AA habitats in Greenland, Iceland, the Alps, Scandinavia, Alaska, and the Rocky Mountains. A complex with a wide distribution in grasslands, dunes, lawns, swamps, burns, and woodlands. August in the alpine.

Notes This complex is well known from lower elevations. Easily recognized by the bright yellow cap, gills, and long stem, which all stain black; whether the alpine version is a separate species is yet to be determined.

- Cap pointed, yellow, viscid, turning black
- Gills yellow, turning black
- Stem, long and thin, yellow, turning black
- Spores white
- In open grasslands or alpine vegetation

Reference Boertmann 1995. CLC 3826b.

Hygrocybe* cf. *punicea (Fr.) P. Kumm.

Cap 2.5–4 cm across, convex or slightly conic-convex, or with indistinct umbo, deep red to orange-red, fading from center out to pale yellowish orange or remaining red, smooth, waxy-viscid; margin turned down. **Gills** narrowly attached, thick, well separated, at first pale orange-cream, then with reddish or yellowish tones. **Stem** 2–5 cm × 0.8–1.2 cm, robust, equal but tapering to a point at base, dry, streaked with yellow, orange, and red tones, pale yellow to white at base. **Flesh** pale yellow, splitting in stem. **Odor** absent. **Spores** white, 10 × 5 µm, ellipsoid, smooth.

Ecology and Distribution Scattered in groups in grass; in the alpine zone on the Beartooth Plateau and expected at lower elevations. Known from Iceland and the Alps in grasslands and dunes late in the field season. September in the alpine.

Notes This robust species with a bright red viscid cap and thick, dry stem stands out in alpine grasslands. Also reported from the eastern and western United States in woods. Other smaller, brightly colored Hygrocybes are also possible, but rare in the Rocky Mountain alpine.

- Cap robust, orange red, waxy-viscid
- Gills pale orange-red
- Stem robust, orange-red, dry
- Spores white
- In open grasslands; late fall, September

References Boertmann 1995; Siegel and Schwartz 2016. CLC 2300.

Lepiota erminea (Fr.) P. Kumm.

Cap 4–5 cm wide, convex but flat or depressed on top, whitish cream, finely hairy-fibrillose (hand lens), with bumps in center; margin with bits of veil tissue, slightly ribbed. **Ring** as bits of tissue on cap margin and/or stem. **Gills** free, whitish cream, rather broad, thickish; edges floccose, graying. **Stem** 3.5–4.0 × 0.8–1 cm, white, striate above, with fine floccules below ring. **Flesh** whitish. **Odor** absent. **Taste** slightly bitter. **Spores** white, spores 12–15(–18) × 5 µm, long ellipsoid, smooth, somewhat dextrinoid.

Ecology and Distribution On Independence Pass, Colorado, at 3,660 m, in grass. Typically, a species of lower elevation grasslands, but noted in alpine habitats in the Italian, French, Austrian, and Swiss Alps as *L. alba*. August.

Notes A pure white *Lepiota* that could be mistaken for a small *Agaricus* except for the white gills and spores. Considered the same as *L. alba* (Bres.) Sacc., which is reported from alpine habitats.

- Cap white, finely fibrillose
- Gills cream, free
- Stem white, rough below ring zone
- Spores white
- In grass

Reference Knudsen and Vesterholt 2008. CLC 1759, CLC 2405b (above).

Lepiota favrei Kühner ex Bon

Cap 1.5–3.5 cm wide, hemispherical, convex, with low umbo, at first totally covered with cinnamon fibrils that separate to become fibrillose brownish scales concentrically scattered on a whitish cream background, with brownish center; margin with whitish veil, scalloped. **Gills** free, broad, crowded, white; edges fimbriate. **Stem** 1.5–2.5 × 0.3–0.6 cm, equal, cream, smooth above the **ring** of pinkish floccules, below with pinkish floccules to base. **Flesh** whitish cream. **Odor** not distinctive. **Spores** white, 8(–10) × 4–5 µm, ellipsoid, dextrinoid.

Ecology and Distribution In clusters on the ground in *Dryas* and dwarf willow in the alpine (but not mycorrhizal). Recorded from the Beartooth Plateau, the Alps, and Greenland. August.

Notes At least eight *Lepiota* species are reported from the alpine. There is some confusion as to the name for this one; we prefer to use the name *L. favrei* after the father of alpine mycology—Jules Favre. *L. pseudolilacea* is possibly the same. It can be mistaken for a *Cystoderma,* but the gills are free.

- Cap convex, covered with pink granules or brownish scales
- Gills white, free
- Stem with pink floccules below membranous ring
- Spores white
- Often in *Dryas*

Reference Peintner 1999. CLC 3523, sequenced.

Lepista irina var. *montana* Bon

Cap 2–8(!) cm wide, convex but flat in center or with indistinct umbo, becoming irregularly lobed, white to buff, smooth, greasy; margin with white hoary coating, rolled under at first then turned down. **Gills** adnate to subdecurrent, narrow, cream to buff; edges concolorous, finely crenate in age. **Stem** 2–5 × 0.6–2 cm, at first gradually larger toward base, then more equal, dingy cream, with a hoary coating at first, then striate, at base white mycelioid. **Flesh** white, solid. **Odor** sweet, of perfume, flowery. **Spores** cream, 7–10 × 5–7 um, ellipsoid, rough in cotton blue.

Ecology and Distribution Forming rings in alpine grasslands on the Beartooth Plateau (Montana), reported in the Alps and possibly Greenland. August.

Notes Recent molecular data groups *L. irina* in the genus *Collybia,* however, we retain the alpine variety as a *Lepista* until further information is available. *Lepista irina* proper has smaller, narrower spores (6–7 × 3.5–4.5 um), and a different ecology. Both have an aromatic odor. The similar-looking, closely related, *Leucocalocybe mongolicum* and *Lepista panaeolus* also occur at high altitudes in Wyoming and China.

- Cap large, whitish buff, smooth, greasy
- Gills cream to buff
- Stem stout, whitish
- Odor sweet, of perfume
- Spores white, rough
- In grasslands, cespitose and/or in rings

References Armada et al. 2024; Jamoni 2008; Bon 1985; Yu et al. 2011; He et al. 2023. CLC 3551, CLC 3397.

Lichenomphalia alpina (Britzelm.) Redhead, Lutzoni, Moncalvo & Vilgalys

Cap 0.5–1.5 cm across, shallowly convex with sunken or flat center, with turned-down margin, bright egg yolk yellow, smooth, waxy; margin scalloped. **Gills** decurrent, triangular, well spaced, few-in-number, bright yellow. **Stem** 0.3–0.4 cm × 0.1 cm, thin, short, equal, smooth, bright yellow, waxy. **Flesh** yellow. **Odor** absent. **Spores** white, 8–14 × 4–6 µm, ellipsoid, smooth.

Ecology and Distribution Scattered on bare soil among minute mosses and green thalli, in alpine habitats in Alaska and on the Beartooth Plateau, also known from the Arctic, including Greenland, Iceland, Svalbard, and the alpine in Europe. August.

Notes This small mushroom-producing lichen appears restricted to cold environments such as the alpine and Arctic. The similar, *L. hudsoniana* is pale yellow with a whitish stipe, occurs in similar habitats, and is reported from Rankin Inlet in Canada. Both can be confused with small Hygrocybes.

- Cap small, bright vivid yellow, waxy
- Gills bright yellow, decurrent
- Stem, small, bright yellow
- Spores white
- On open soil among mosses & lichens

Reference Knudsen and Vesterholt 2008. CLC 3812.

Melanoleuca cognata (Fr.) Konrad & Maubl.

Cap 7–11 cm wide, broadly convex with low round umbo, becoming flat or depressed with uplifted margin, buff with orange cast, orange-cream, smooth, greasy; margin with darker rim. **Gills** deeply adnexed, crowded, becoming broad, white, then cream; edges concolorous. **Stem** 5–8 × 1–1.5 cm, tall, equal, or enlarged at the base, pale orange-cream, with striate-twisted surface. **Flesh** white in cap, pale orange in stem. **Odor** not distinct, possibly faintly fruity. **Spores** cream, 8–9 × 5–6 μm, ellipsoid, with amyloid warts; cystidia pointed, incrusted, base swollen.

Ecology and Distribution Scattered to solitary in open grassy meadows at high elevations (up to 12,300'/3,650 m) in the alpine or krummholz in the Rocky Mountains. Also known from alpine grasslands in Europe and Arctic areas of Greenland. Found at lower elevations as well. August.

Notes This large orange-buff species is common in the alpine. The genus can be recognized in the field by the striate-twisted stem, flat cap, and amyloid warted spores. This may be the same as *Melanoleuca borealis*. *M. exscissa* is less orange.

- Cap large, almost flat or dished, orange-buff
- Gills crowded, cream
- Stem long and tall, pale orange, twisted-striate
- Spores cream, with amyloid warts
- In alpine meadows or grasslands, often in rings

References Gillman and Miller 1977; Kühner 1978. CLC 2272, Genbank MH379762.

Melanoleuca exscissa (Fr.) Singer

Cap 4–14 cm wide, flat with round umbo, depressed around umbo, white, cream, buff, light brown, pale yellow-buff; margin turned down, cuticle overhangs gills. **Gills** adnexed to subdecurrent, narrow, crowded, cream with a gray cast. **Stem** 6–12 × 0.8–2.0 cm, tall, straight, equal or somewhat larger at base, dingy cream, pale brown, totally pruinose when young, then longitudinally fibrous; white myceloid at base. **Flesh** white; stem stuffed. **Odor** slightly mealy or of licorice? **Spores** white 8–9 × 5 µm, ellipsoid with amyloid warts; cystidia pointed with crystals, swollen at base, rare.

Ecology and Distribution In high elevation open grasslands, often forming large rings in fall. Occurring in the Rocky Mountains of Colorado, Wyoming, and Montana. Well known from Europe in grassy habitats at lower elevations, and in parks and gardens.

Notes The ITS sequence of our specimens matches the epitype for *M. exscissa* 100%; the cap of this species varies from gray to whitish. *M. substrictipes* and *M. subalpina* have white to cream caps and should also be considered. *M. cognata* caps have an orange cast.

- Cap large, flat with round umbo, cream, buff
- Gills cream, crowded
- Stem, tall, cream, pale brown, twisted-striate
- Spores white, with amyloid warts
- In alpine grasslands

References Antonin et al. 2017; Sanchez-Garcia et al. 2013; Gillman and Miller 1977. CLC 3549, Genbank MH379767.

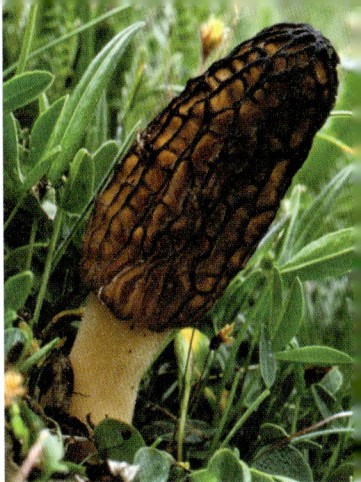

Morchella norvegiensis Jacquet. ex R. Kristiansen

Cap 2–3 cm across × 2–4 cm high, conical with mostly rounded apex, covered with longitudinally arranged pits; pit interior yellow-brown; ridges brown then black; edge of cap attached to stem. **Stem** 2–3 cm high × 1–2 cm wide, equal or narrower in middle, cream, buff, covered with granular bumps, hollow, indented where attached to cap. **Flesh** brittle, cream to buff, bumpy inside; fruiting body hollow. **Odor** not distinct. **Spores** 21–25 × 15–18 μm, ellipsoid, smooth.

Ecology and Distribution In *Dryas octopetala,* dwarf willows or vegetation in open mesic areas of the alpine on the Beartooth Plateau in Montana and Loveland Pass in Colorado. Known from Norway and Northern Europe, the Pacific Northwest, and the Rocky Mountains from sub-alpine habitats, primarily conifer and deciduous forests. August in the alpine.

Notes Sequences match *M. norvegiensis.* Larger specimens from adjacent subalpine sites may be the same. A rare species in the alpine, it should not be considered for food but conserved. Perhaps the world's highest morel, one collection was found at 12,000 ft. (3,660 m).

- Cap oblong, covered with pits and dark ridges
- Stem cream, bumpy
- Cap and stem hollow
- In open mesic areas, with alpine vegetation
- Sometimes in *Dryas*

Reference Clowez and Moreau 2021. CLC 1784, sequenced.

Mycena* cf. *diosma Krieglst. & Schwöbel.

Cap 1–3.5 cm wide, shallow convex, smooth, light to medium brown in center, outward mauve color (reddish with purplish hues), with a concentric ring, striate and lilac at the edge. **Gills** deeply adnate, broad, thick, well separated, pale lavender. **Stem** 3–6(!) × 0.4–0.6 cm, equal, long (in moss), pale to dark lavender, silky smooth. **Flesh** lavender with yellow hues; stem hollow. **Odor** not distinct or radish-like. **Spores** white, 6–9 × 4.5 μm, long ellipsoid to dacrymoid, smooth.

Ecology and Distribution *Mycena diosma* occurs occasionally in Arctic-alpine habitats in Europe and is reported here from alpine Alaska. August.

Notes *Mycena diosma* is phylogenetically and morphologically close to *Mycena pura,* which is a variable complex that needs to be sorted out; apparently ITS is not sufficient. *M. diosma* has a more striate-banded cap, and usually a strong fragrant odor that was lacking in ours; it is also reported from lower elevations.

- Cap brown with purple hue, zoned, striate
- Gills and stem lavender
- Spores white
- Odor should be radish-like
- In grass and alpine vegetation

Reference Harder et al. 2013. CLC 3814.

Mycena pura (Pers.) P. Kumm. complex (nano form)

Cap 1–1.5 cm wide, convex, lavender to buff, smooth, silky, indistinctly striate; margin turned in, crenate. **Gills** attached, broad, thickish, pale lavender; edges jagged. **Stem** 1.5–2.5 × 0.1–0.3 cm, equal or a bit larger at base, lavender, whiter at base, smooth but striate. **Odor** absent but should be radishy. **Flesh** pale lavender. **Spores** white, 7–8 × 3–4 µm, long ellipsoid, smooth.

Ecology and Distribution *Mycena pura* is a broadly distributed species complex found above and below treeline. Also known from AA habitats in Europe. Ours is from the Beartooth Plateau (3200 m) in *Dryas*. August.

Notes The ITS sequence places this collection in the *Mycena pura* complex, but additional genes are needed to determine the subgroup. This alpine taxon is unique in its miniature habit, fleshy stem, and ecology in *Dryas*. *M. rosea* is larger with a pink umbonate cap. Bon described *M. pura* var. *luteorosa* from the alpine which is more pinkish yellow.

- Cap small, lavender, smooth
- Gills and stem lavender
- Spores white
- Odor should be radish-like but was absent
- In *Dryas*, for this collection

Reference Harder et al. 2013. CLC 3523, Genbank PQ191441.

Stropharia alpina (M. Lange) M. Lange

Cap 2.5–5 cm wide, convex, pale to deep turquoise mottled with ocher, smooth, sticky-viscid; margin with a rim of darker bluish green or bits of white veil tissue. **Gills** deeply indented, pale to medium gray, browner with spores, broadest in middle; margin floccose, sometimes crenate. **Ring** present as slight tissue on stem or cap edge. **Stem** 4.5 × 0.6 cm, equal or tapering toward base, dry, pale gray to pale turquoise (bluish green), smooth above and cottony floccose below ring zone. **Flesh** pale turquoise. **Odor** not distinct. **Spores** purple-brown, 8–10 × 5–6 µm, ellipsoid, smooth, with small germ pore; chrysocystidia absent.

Ecology and Distribution In the Rocky Mountains, on rocky ledges on the Beartooth Plateau (Montana) and in grasslands on Independence Pass, Colorado. Also in alpine and Arctic areas of Norway, Iceland, and Greenland. August.

Notes Molecularly closest to *S. cyanea*, but not that close. This species is reported from numerous Arctic and alpine environments. The beautiful turquoise color and purple black spores are diagnostic. It should be compared to subalpine turquoise species.

- Cap convex, pale to dark turquoise
- Gills gray to purple-brown when mature
- Stem pale turquoise with slight ring tissue
- Spores purple-brown/black with germ pore
- In grasslands, healthlands, and on rocky ledges

Reference Knudsen and Vesterholt 2008/2012. CLC 1106 sequenced; CLC 1758, Genbank PQ196639.

2

Puffballs in Meadows and Grasslands

Puffballs can be found scattered across grasslands and meadows in the open alpine, sometimes arranged in rings or arcs. The rings are a result of underground myelium growing out in all directions from where the initial spores landed. The open tundra-like landscape offers few barriers to expansion, and rings can become huge in the alpine. Nitrogen released by these decomposers produces the dark green color of the vegetation ring; plant species are often different inside or outside of the ring. Some *Agaricus* species form rings that look similar to puffball rings from a distance, which is interesting as these ecologically similar, but morphologically distinct fungi are evolutionarily related. Puffballs were once grouped in Gasteromycetes (stomach fungi) because their spores are internal, but this is an artificial group.

Some of the largest fungi in the alpine are puffballs. The rare Giant Western Puffball (*Calvatia booniana*) can be over a foot across, but most puffballs range from baseball-size down to marble- size (*Bovista limosa*). Puffballs have several mechanisms for releasing the spores. In some, the external layer (peridium) cracks open and weathers away (*Calvatia*-type), sometimes leaving a cup. In the tumbling puffballs (*Bovista*), the outer layer deteriorates, revealing an inner papery layer that detaches from the soil and blows around, spreading spores. The smaller puffballs (*Lydoperdon*-type) have a pore (ostiole) that puffs out spores when hit by raindrops, hence the name puffball. Internal capillitium (skeletal hyphae) keeps puffballs from collapsing in on themselves.

Some *Calvatia, Bovista,* and *Lycoperdon* appear restricted to Arctic and alpine habitats, while others also occur in lower or warmer regions.

Decomposers

Key to Puffballs in Meadows and Grasslands

1. Sterile base absent; spores *usually* with long tails; capillitium wide 2
1. Sterile base present; spore tails absent (short in one); capillitium narrow 5

2. Fruitbody small (5–15 mm) with a fimbriate pore **Bovista limosa**
2. Fruitbody larger, detaching from soil; inner layer papery, with a pore 3

Puffball in the alpine.

3. Inner layer lead gray; spores smooth, ovoid with long tails **Bovista plumbea**

3. Inner layer coppery; spores verrucose, round; tails long or short 4

4. Spore tails very short ≤ 2 um; spores 3.5–4.5 μm **Bovista pila**

4. Spore tails longer 8–12 μm; spores 4–5.5 μm **Bovista nigrescens** (see *B. pila*)

5. Pore present to release spores; stem-like base present or not 6

5. Outer layer breaking up to release spores, leaving a cup-like base 7

6. Surface with fine tufts of spines; spores spiny **Lycoperdon frigidum**

6. Surface with dense pyramidal spines; spores smooth **Lycoperdon norvegicum** (see *L. frigidum*)

7. Top part a mosaic of plates that disintegrate, leaving a vase-like cup with patterned exterior; spores smooth
Lycoperdon utriforme
7. Top part a fine pattern of spines or warts; spores verrucose
8

8. Outer layer thick, with warts or spines; spores verrucose
Lycoperdon cretaceum
8. Outer layer thin, with fine spines; spores sparsely verrucose
Lycoperdon turneri

Bovista limosa Rostr.

Fruiting body tiny, 5–10 mm across, round to depressed ellipsoid. **Outer layer** at first pure white, yellowing in age, smooth or with minute spines or flakes at base; at maturity **inner layer** is papery, reddish brown, brown, with scurfy texture. **Pore** (ostiole) protruding from surface like a beak. **Interior** (gleba) dark brown at maturity, powdery. **Sterile base** absent. **Odor** not distinct. **Spores** brown 4–5 μm, round, thick-walled, very slightly verrucose, with a tail several microns long.

Ecology and Distribution Scattered among mosses or on bare soil; occurring in the Rocky Mountain alpine from Colorado to Montana and reaching into the Canadian Arctic. Known from Northern Scandinavia and Greenland; apparently restricted to cold climates. July.

Notes Regarded as the world's smallest puffball, this species has been misidentified as *Lycoperdon pusillum* or *L. echinulla*. Molecular data confirms it occurs mostly in cold climates.

- Fruiting body tiny, round, white, almost smooth
- With fimbriate pore
- Turning papery reddish brown, brown
- Spores with a tail
- Among mosses or on open soil

References Kasuya 2010; Jeppson 2018; Larsson et al. 2009. CLC 1814.

Bovista pila Berk. & M.A. Curtis

Fruiting body 3–6 cm wide × 2–3 cm high, ovoid, slightly flattened. **Outer layer** (exoperidium) white, bruising pinkish, delicately textured, flaking off in patches. **Inner layer** (endoperidium) papery, coppery, bronze, shiny, smooth, with darker areas, detaching from substrate. **Pore** irregular, as a rip or tear. **Interior** (gleba) dark brown, purplish brown, powdery. **Sterile base** absent. **Odor** not distinct. **Spores** deep brown, round, 3.5–4.5 µm, almost smooth, with very short tail ≤ 2 µm or absent; capillitium very wide –13 µm, branching, tapering to a point.

Ecology and Distribution This North American species has been found in the low alpine (3,200 m) in Wyoming and Montana. It is also widespread in temperate zone pastures, open woodlands, and rangeland at lower elevations. July, August.

Notes *Bovista pila* is not a typical alpine or Arctic species. A more common species in cold environments (particularly in Europe) is *B. nigrescens,* which has spores with longer tails (–12 µm); *B. nigrescens* is also reported from the Beartooth Plateau.

- Fruiting body ovoid, white, textured
- Outer layer becoming areolate, flaking off in patches
- Inner layer papery, coppery brown, detaching from soil
- Spores ovoid with short tails; capillitium very wide
- In open grasslands

References Kasuya 2010; Jalink 2010. CLC 1290.

Bovista plumbea Pers.

Fruiting body 2–4 cm wide, round to ovoid. **Outer layer** (exoperidium) white at first, slightly patterned, becoming areolate with flat plates separated by dark brown fissures; plates flaking off. **Inner layer** (endoperidium) papery, lead gray to blackish; detaching and becoming wind-blown. **Pore** present for spore release. **Interior** (gleba) brown, olivaceous brown, yellow-brown, powdery. **Sterile base** lacking. **Spores** 5–6(–7) × 5–6 µm, yellow-brown, ovoid, smooth, with a 10–15 µm long tail. **Capillitium** a tangle of wide hyphae –18 µm, with thick walls –3 µm.

Ecology and Distribution Widely distributed in grasslands and rangelands from low to high elevations. Less frequent at higher elevations but common in the alpine on the Beartooth Plateau and known from high elevations of northern Norway, Iceland, and the Faeroe Islands.

Notes Unlike *B. nigrescens*, the outer layer of *B. plumbea* flakes off in sections. In young stages, it could be mistaken for a *Lycoperdon,* but fruiting bodies lack a sterile base, and spores have long tails.

- Fruiting body ovoid, white with slight pattern, detaching from soil
- Outer layer becoming areolate, cracking into plates
- Inner layer papery, lead gray
- Spores ovoid with long tails; capillitium very wide
- In open grasslands

References Jeppson 2018; Kasuya 2010. CLC 3175.

Lycoperdon cretaceum Berk.

Fruiting body 4–6 cm wide × 5 cm high, round to ovoid, can be pear-shaped, puckered at base. **Outer layer** (exoperidium) thick, white, with flat to distinct pyramidal warts in top part, and spines in lower part, smoother in age; spines can be peeled off. **Inner layer** (endoperidium) white, becoming gray, smooth; top part breaking up and lower part persistent as a vase-shaped cup. **Interior** (gleba) dark brown, olive-brown, powdery. **Sterile base** locular; can be broadly stem-like. **Odor** not distinct. **Spores** 5–6 µm, round, verrucose; capillitium narrow, fragile, breaking up; with slits.

Ecology and Distribution Common in Arctic-alpine habitats, mostly in open grasslands. Reported from Iceland, Fennoscandia, the Alps, Siberia, and now the Rocky Mountains above treeline.

Notes Also known as *Calvatia cretacea*. Highly variable in morphology and ornamentation, but with a thick outer skin that peels and a lower part that persists as a cup, similar to *L. utriforme* and *L. turneri* (thin outer layer), which have smooth or less spiny spores, respectively.

- Fruiting body round, ovoid, pear-shaped, white
- With flat pyramidal warts on top, and spines below
- Outer layer thick, peeling
- Inner layer grayish, persisting as a cup
- Spores verrucose; capillitium fragile
- In open alpine vegetation, including grasslands

References Jeppson 2018; Kasuya 2010. CLC 3555.

Lycoperdon frigidum Demoulin

Fruiting body 2.5–4.0 cm wide, 3–5 um high including base, pear-shaped and puckered at base. **Outer layer** (exoperidium) whitish or very pale grayish brown, with fine convergent warts, spines, or granules, which fall off; underside smooth or with a few spines. **Inner layer** (endoperidium) areolate golden brown, papery with granules. **Sterile base** present, stem-like, spongy-porous. **Pore** round to lacerate. **Interior** (gleba) brown, powdery. **Odor** not distinct. **Spores** 4–6 μm, round, moderately spiny/verrucose, most lacking tails. Capillitium 3–6 μm wide.

Ecology and Distribution Occurring on the Beartooth Plateau in *Dryas* and willows; an Arctic-alpine species that is abundant in open heath and grasslands. It is known from Arctic Canada, Greenland, Iceland, Fennoscandia, Siberia, and the Alps. August.

Notes Similar to the Arctic-alpine *L. niveum*, which has more conspicuous spines, and smaller spores, as does *L. molle*. *L. norvegicum*, also reported from the Beartooth Plateau, has smooth spores. *L. frigidum* is only recently reported from the lower forty-eight states from our alpine sites.

- Fruiting body pear-shaped, white, with fine spines/warts
- Becoming papery, golden brown in age; top areolate
- With pore when mature
- Interior brown, powdery
- Spores round, spiny
- In open grasslands or in *Dryas* or willows

References Jeppson 2018; Kasuya 2010. CLC 3559.

Lycoperdon turneri Ellis & Everh.

Fruiting body 4–8 cm across by 3–6 cm high; round, flattened ovate, or sometimes pear-shaped; thin-skinned; in age forming vase-shaped open cup. **Outer layer** (exoperidium) dingy white to cream color; surface pattern of very fine spines in clusters; spines eroding, developing a patterned, cracked appearance. **Inner layer** (endoperidium) grayish brown, brown. **Interior** (gleba) white, then yellow-brown or very dark brown, powdery. **Pore** absent. **Sterile base** present but can be minimal. **Odor** somewhat unpleasant. **Spores** yellowish brown, 4.5–5.5 μm, round, moderately verrucose, low warts; capillitium narrow, fragile, breaking up.

Ecology and Distribution Scattered in open grasslands at high elevations. One of the most common puffballs in Arctic-alpine habitats, it has been reported from Alaska, Montana, and Canada in North America and has a circumpolar distribution. August.

Notes Also known as *Calvatia turneri*. *Lycoperdon cretaceum* is similar but has coarser ornamentation, a thicker skin that peels off, and larger spores. It is also found in the Rocky Mountain alpine.

- Fruiting body, flat ovoid, white with fine spines
- Outer layer thin, base eventually forming a wide cup
- Interior dark brown, powdery
- Spores round, somewhat spiny
- Capillitium narrow, fragile
- In open alpine grasslands and meadows

References Jeppson 2018; Kasuya 2010. CLC 2315.

Lycoperdon (Calvatia) utriforme Bull.

Fruiting bodies 5–7 cm wide, 4–8 cm high, turbinate with elliptical top part and narrower sterile base, which can be indistinct. **Outer layer** (exoperidium) white, cream, pale brown, fissuring into plates, creating an areolate pattern, smoother in lower half, opening by breaking up of peridial wall; plates can have pyramidal warts; base eventually remaining as a vase or cup of spores. **Inner layer** (endoperidium) thin, brown with silvery hue, papery. **Interior** (gleba) powdery, dark brown. **Sterile base** present. **Odor** not distinct. **Spores** brown, 4.5–5.5 µm, round, smooth, thick-walled. **Capillitium** with pores and slits, septa rare.

Ecology and Distribution Common on the Beartooth Plateau in alpine vegetation and known from low elevations in Iceland. Reported in Wyoming and Colorado as *Calvatia bovista* and from Utah as *Calvatia tatrensis.* Widespread but not typically from alpine or Arctic habitats. July to August.

Notes Called the "mosaic puffball." Once placed in genera *Handkea, Calvatia,* or *Bovistella,* the molecular data places this species in genus *Lycoperdon.* The smooth spores and capillitium with pores and slits are the important features. The base can persist until the second season.

- Fruiting body top-shaped, cream, pale brown
- Outer layer cracking to form plates, becoming areolate
- Spores brown, smooth; capillitium with slits
- In open alpine grasslands and in *Dryas*

References Jeppson 2018; Kasuya 2010; Jalink 2010. CLC 3395, CLC 2717.

Lycoperdon utriforme CLC 3395.

Giant Western Puffball *Calvatia booniana*. Rare in the North American alpine but possible.

Numerous minute mushrooms can be found nestled in mosses and on liverworts or spread out on open wet soil in alpine and Arctic situations. They fruit in seeps, below melting snowbanks, along streams and around pooled water. Mosses and lichens are prolific as ground cover in cold climates. Two of the most common and diverse moss-loving genera, *Mycena* and *Galerina,* often fruit on dead brown parts of mosses, suggesting that they are decomposers. Indeed, many have enzymes for decomposition of dead plant material. The same may be true for *Hypholoma* and *Omphalina* species plus those of *Loreleia* (which fruit on liverworts). Some moss-associates have a preference for *Sphagnum*

and *Dicranum* mosses. A few of the tiny moss mushrooms are parasitic, including certain species of *Rickenella* and *Deconica*. Arrhenias often fruit on wet soil, although they are also found on willow branches and in moss. Some Arrhenias are easily recognized by their reduced, veinlike gills. The small, bright yellow basidiolichen *Lichenomphalia* is treated in the Meadow Mushroom section.

Mycena and *Galerina* species tend to have bell-shaped caps, while other moss mushrooms have funnel-shaped caps, including species of *Rickenella, Arrhenia*, and *Loreleia*. All of these have white spores except the Galerinas, which have brown spores. *Hypholoma* and *Deconica* species have blackish spores when mature. However, gills can be pale in young fruiting bodies of all these genera, making spore color difficult to determine. Carefully check the top part of the stem with a hand lens for dropped spore color.

Key to Mushrooms on Mosses, Liverworts, or Open Soil

1. Stem attached to side of cap or absent 2
1. Stem present and central 4

STEM ATTACHED TO SIDE OF CAP

2. Gills as normal blades; cap tiny, 7–9 mm, buff; stem tiny, hirsute **Arrhenia sp.**
2. Gills reduced to veins; cap 6–20 mm, grayish brown; stem present or absent 3

3. With distinct stem; cap spoon-shaped **Arrhenia auriscalpium**
3. Stem not distinct or absent; cap fan-shaped, lobed **Arrhenia lobata**

SPORES WHITE; GILLS DECURRENT OR ALMOST FREE

4. Spores white; gills decurrent or almost free 5
4. Spores dark: brown or purple-black; gills attached 11

5 Cap bell-shaped, striate; gills almost free 6

5. Cap funnel-shaped; gills decurrent 7

6. Cap yellow, olive-yellow **Mycena citrinomarginata**

6. Cap very pale brown **Mycena cf. pasvikensis** (also see *Rhizomarasmius*)

7. Cap orange 8

7. Cap with duller colors; brown, blackish brown 10

8. In moss; cap orange; stem thin, rubbery, pale orange **Rickenella fibula**

8. On liverworts; stem fragile or fleshy 9

9. Cap orange to pale orange; gills pale yellow, separated **Loreleia marchantiae**

9. Cap bright orange; gills white, more numerous **Loreleia postii**

10. Cap light brown, striate; stem long, thin, fragile **Omphalina rivulicola**

10. Cap brown, blackish brown, striate; stem short **Arrhenia velutipes**

SPORES DARK; GILLS ATTACHED

11. Spores brown, orange-brown; cap mostly orange-brown 12

11. Spores purple-black; cap olive-brown, brown, red-brown, yellow-brown 16

12. Cap fleshy, convex, not striate; stem with obvious ring 13

12. Cap more fragile, bell-shaped, striate; stem without obvious ring 14

13. Cap orange-brown; stem longer than cap width; spores smooth **Galerina pseudomycenopsis**

13. Cap orange-brown; stem shorter than cap width; spores verrucose **Galerina marginata**

Mossy alpine habitat.

14. Cap yellowish to orange; gills orange, well separated; stem whitish, top part pruinose; cystidia tibiiform; spores 12–13 µm, lightly verrucose **Galerina clavata** (& *G. subclavata*)
14. Cap yellowish to orange; gills orange; stem pale, totally pruinose; cystidia fusiform-ventricose; spores 10–12 µm, distinctly verrucose 15

15. Cap with pileocystidia **Galerina atkinsoniana**
15. Cap without pileocystidia **Galerina vittiformis**

16. Cap smooth, lubricous, olive-yellow, not striate **Hypholoma**
16. Cap red-brown, brown, yellow-brown, striate **Deconica montana**

Other small mushrooms not usually in moss: *Hygrocybe* (yellow, orange, red), *Lichenomphalia* (yellow), *Laccaria* (orange), *Gymnopus* (brown), *Mycena pura* (lavender).

Arrhenia auriscalpium (Fr.) Fr.

Cap very tiny, 3–6(–15) mm across; erect, spoon-shaped, thin-fleshed, pale brown to gray-brown, somewhat translucent, often with a lobed margin that is curled over. **Gills** reduced to a few shallow ridges that branch toward the margin, color same as cap. **Stem** 2–4 mm long × 1–2 mm wide; smooth or minutely velvety; pale brown to gray-brown. **Spores** white, 8.5–9.5 × 5–5.5 μm, tear-drop shaped, ovoid, smooth, not amyloid.

Ecology and Distribution In true alpine habitats at high elevations and in the Arctic near sea level. Occurring on dark, open, organic, sun-exposed soil near minute mosses. Reported from Loveland and Independence Passes in Colorado, also in Alaska and Canada in North America. In AA habitats in Greenland, Iceland, the Alps, Scandinavia, Svalbard, Russia, and Alaska. Rare or rarely noticed. August.

Notes The similar and more common *Arrhenia lobata* occurs in the alpine as well as subalpine habitats; it is usually attached to mosses, and it also has reduced ridges instead of gills, but the cap is more lobed, and it lacks a distinct stem.

- Cap very tiny, spoon-shaped, gray-brown
- Gills reduced to a few vein-like ridges
- Stem attached to one side of cap
- On open, dark, organic soil or in moss; rare

References Cripps and Horak 2006; Cripps et al. 2016. CLC 3803.

Arrhenia lobata (Pers.) Kühner & Lamoure ex Redhead

Cap up to 2 cm across, irregularly fan-shaped, lobed, thin-fleshed, translucent, grayish brown, brown, smooth or wrinkled, laterally attached to mosses; margin wavy. **Gills** present as branching veins, paler than cap. **Stem** absent. **Flesh** thin, rubbery soft, flexible. **Odor** not apparent. **Spores** white, ellipsoid, 8–10 × 4.5–6.5 µm, sometimes longer.

Ecology and Distribution Scattered or in clusters in mosses in wet areas along streams, occasionally on willow twigs. Found in high-elevation subalpine forests or alpine habitats. Reported from many AA habitats, including in Europe, Greenland, Canada, Alaska, the Rocky Mountains, and Antarctica. August.

Notes Molecular work has shown that *Arrhenia* species with veins instead of gills are in the waxcap family Hygrophoraceae. They can be prolific on mosses along alpine streams but absent when high runoff scours the vegetation. *A. auriscalpium* has a distinct stem and occurs on soil.

- Fruiting body fan-shaped, lobed, grayish brown, translucent
- Gills reduced to branching veins
- Stem absent
- Attached to mosses or twigs, often clustered

Reference Redhead 1984. CLC 1791.

Arrhenia sp.

Cap 7–9 mm in diameter, pleurotoid, clam-shaped (semi-orbicular), pale gray, grayish buff, with darker but faint striations, white hairy near stem attachment point; margin a bit white. **Gills** narrowly attached, blade-like, sharp, somewhat thick, L = 10, but 2nd or 3rd tier of lamellulae also present, not anastomosing. **Stem** laterally attached, inserted on side of cap and underneath, 1–2 mm long × 1 mm wide, finely white hairy overall. **Spores** not measured, scant material available.

Ecology and Distribution Attached to the side of an animal hole with gills facing down, in an alpine meadow near mosses in the San Juan Mountains, Colorado. August.

Notes Molecularly closest to collections named *Arrhenia leucotricha* and *Arrhenia (Pleurotellus) acerosa*, but not a close match. The small clamshell cap, well-formed gills and lateral white, fuzzy stem are the main features.

- Cap clam-shell shaped, grayish buff, tiny
- Gills well formed, blade-like
- Stem attached to one side, white, pubescent
- In animal hole, in alpine meadow

Reference Voitk et al. 2020. CLC 1743, Genbank PQ165957.

Arrhenia velutipes (P.D. Orton) Redhead, Lutzoni, Moncalvo & Vilgalys

Cap 1–2 cm wide, deeply funnel-shaped, blackish brown when fresh, with black striations on brown background, hygrophanous, drying lighter brown, a bit rough, opaque; margin sulcate or not. **Gills** decurrent, medium to pale brown, well spaced, about 16 reach the stem; edges darker or not. **Stem** 1–1.7 × 0.15–0.3 cm, equal, a bit curved, blackish brown, gray-brown, white at base, velvety, more obvious in lower half. **Flesh** gray. **Odor** not distinct. **Spores** white, 7–8 × 4–6 μm, ellipsoid, teardrop shaped, smooth.

Ecology and Distribution In pioneering situations on open soil, mud flats, and near old mine sites with disturbed soil; common in the Rocky Mountain alpine. Also reported from the Alps, Fennoscandia, Scotland, Iceland, Greenland, Russia, and Canada. Circumpolar. August.

Notes The Arrhenias in AA habitats are difficult to distinguish from each other. This species has a velvety stem, which is not always obvious, but hairs can be observed with a hand lens or microscope; some are encrusted and thick walled. One of the few Arrhenias on soil.

- Cap small, funnel shaped, black or brown, striate
- Gills decurrent, brown, well separated
- Stem dark blackish brown, velvety when fresh
- Spores white
- On open soil, mud flats, disturbed areas

References Gulden and Jenssen 1988; Knudsen and Vesterholt 2008. CLC 1418 (above); CLC 3605, Genbank PQ167989.

Deconica montana (Pers.) P.D. Orton

Cap 0.5–1.5 cm wide, convex, some with slight umbo, dark red-brown, paling to yellow-brown, viscid when wet, to dry; margin pleated-striate with white edge. **Gills** adnate to short decurrent, well separated, brown, red-brown, blackish, brown. **Stem** 1–4 × 0.1–0.2 cm, equal, reddish brown, yellow-brown, with a few hairs. **Odor** not distinct. **Spores** purple-black, grayish black, 6–9 × 4–6 µm, ellipsoid, smooth, with germ pore; with cystidia.

Ecology and Distribution On dead or live mosses, often *Polytrichum* in Arctic-alpine habitats; also found in temperate forests. Reported in AA habitats in Fennoscandia, the Alps, Greenland, and Canada, as well as Alaska and the Rocky Mountains. July-August.

Notes Previously known as *Psilocybe montana*. The other species commonly reported from Arctic-alpine habitats, *D. chionophila,* may or may not be the same. It is thought to be a moss parasite, while *D. montana* may be more of a decomposer.

- Cap small, convex, pleated, red-brown, yellow-brown
- Gills almost decurrent, brown, red-brown, black-brown
- Stem paler brown
- Spore print purple black, grayish black, with germ pore
- Scattered in moss patches, often *Polytrichum*

Reference Knudsen and Vesterholt 2008.

Note on Galerina

Galerinas are among the tiniest and most fragile and fleeting mushrooms in alpine and Arctic habitats—and they are diverse. They typically associate with moss in wet areas along streams, seeps, and pond edges. Some species occur with sphagnum mosses, while others prefer different kinds of mosses. Recognition of the moss host can help with identification of the species.

Caps are usually bell-shaped but can be hemispherical or almost flat on some of the larger species (which usually have a ring). Cap colors range from pale yellow, yellow-brown, orange-brown to red-brown, and the cap margin is often striate. The attached gills are usually in the same color range but paler. Stems are slim and can be very fragile. It helps to notice if the stem is partially or totally pruinose. To do this, avoid touching the stem when collecting. Use a hand lens to observe any cystidia, which sparkle in sunlight.

Microscopic features are important in identifying *Galerina* species. All have brown spores, but they vary from almost smooth in some species to warty (verrucose) in others. Most of the species we describe here have large cystidia, and their shape can be diagnostically important. Over thirty species of *Galerina* have been reported in cold climates such as the alpine and Arctic; some are AA species and others transcend treeline. Here we describe five of them, but we expect more species to be present. Key to species on pg 61.

Galerina atkinsoniana* var. *atkinsoniana A.H. Sm.

Cap 0.5–1 cm wide, bell-shaped, bright orange-brown, darker in center, smooth, striate almost to center, moist, shiny; margin even. **Gills** adnate, almost triangular, well separated, pale orange-brown. **Stem** 1–2 × 0.1 cm, pale orange-brown, darker at base, no veil tissue observed; pruinose to base. **Flesh** darkish. **Odor** not determined. **Spores** 10–12 × 7–8 μm, ovoid to slightly almond-shaped, distinctly verrucose, dextrinoid; most basidia 2-spored; cystidia ampul-shaped; pileocystidia present.

Ecology and Habitat Common in the Rocky Mountains in moist alpine, boreal, and subalpine habitats among mosses. Well known in Arctic and alpine habitats elsewhere, especially in *Polytrichum, Dicranum,* and *Sphagnum.* August.

Notes *Galerina atkinsoniana* is similar to *G. vittiformis* but the former is usually more highly colored (darker orange) and has pileocystidia. Our variety *atkinsoniana* has 2-spored basidia. The ITS sequence is close to that of *G. minima,* which is also small but has 4-spored basidia and smaller spores (unless this is a 2-spored version of *G. minima*).

- Cap orange to yellow, striate, bell-shaped
- Gills paler
- Stem thin, pale, and pruinose
- Spores verrucose; cystidia ampul-shaped; with pileocystidia
- In moss in wet areas

References Armada et al. 2024; Gulden 2010. Smith 1953. Smith 1964. CLC 3902, sequenced.

Galerina clavata (Velen.) Kühner

Cap 0.5–1.6 cm wide, bell-shaped, with rounded apex, caramel, bright orange-brown, more brown in center, slightly pleated-striate at edge, smooth in center, greasy looking, fragile; margin often flared out. **Gills** deeply adnexed, well separated, thickish, pale orange. **Stem** 1.5–3.0 × 0.1 cm, equal, pale orange, whitish, silky-fibrillose, pruinose at least at apex. No veil noted. **Flesh** pale. **Odor** absent. **Spores** 12–13 × 6–8 µm, long, almond-shaped, lightly verrucose; basidia 4-spored; cystidia tibiiform,

Ecology and Distribution Widely distributed in alpine and Arctic habitats but rarely recognized. In the Rockies, found from Colorado to Canada, always in mosses. Known from alpine, Arctic, and temperate areas of Europe. August.

Notes The well-separated, bowed-out orange gills, tibiiform cystidia, and large spores appear distinctive. *G. arctica* has smaller spores; *G. subclavata* has 2-spored basidia.

- Cap pale orange, bell-shaped, striate
- Gills pale orange
- Stem whitish, fragile, top pruinose
- Spores large; cystidia tibiiform
- In mosses

Reference Gulden 2010. CLC 3608.

Galerina cf. *marginata* (Batsch) Kühner

Cap 1.5–3.5 cm, convex, shallow convex with a wavy uplifted margin, deep caramel, hygrophanous drying ocher, smooth, greasy, at margin slightly striate or not; margin with a dark rim. **Gills** adnexed, becoming broad, orange-brown; edges can be dark. **Ring** white, membranous or as white tissue. **Stem** 1.5–3 × 0.2–0.4 mm, equal, cream turning almost blackish, smooth. **Flesh** dark. **Odor** not noted. **Spores** 9–10 × 5 µm, slightly almond-shaped, verrucose, some with tight calyptra; cystidia ampul-shaped.

Ecology and Distribution Reported from the Beartooth Plateau alpine of Wyoming/Montana in wet moss near *S. reticulata*; also likely in the Colorado alpine. More commonly known from subalpine habitats in moss on conifer wood in North America, Europe, and elsewhere. August.

Notes Similar to *G. pseudomycenopsis,* which has smooth spores but is more restricted to AA habitats. Potentially deadly. The name *G. marginata* likely covers a complex of species.

- Cap orange-brown to red-brown, smooth, not or slightly striate
- Gills orange-brown
- Stem pale, turning dark; with white membranous ring
- Spores verrucose
- In mosses or on wood, in the alpine and at lower elevations

Reference Gulden 2010. CLC 3609, sequenced, matching *G. marginata* 99.49%.

Galerina pseudomycenopsis Pilát

Cap 0.3–1.5 cm in diameter, almost hemispherical, then convex, smooth, not translucent, not striate, bay brown, brown, drying ocher from top down, greasy to dry, shiny; margin turned in at first. **Gills** adnate, broadly attached, slightly separated, pale ocher, then cinnamon brown. **Ring** semi-membranous, flaring out, superior, white. **Stem** 1.5–3 × 0.2 cm, long and thin, narrowest at apex, smooth, pale yellow-brown, with a few white fibrils below ring, dark at base to blackish. **Odor** not noted. **Flesh** yellow-brown, blackish at base. **Spores** 10–12 × 6–8 μm, slightly almond-shaped, some long and narrow, almost smooth; cystidia ampul-shaped.

Ecology and Habitat In the Rocky Mountains in moist mossy areas, including iron fens with *Polytrichum*. Reported from Colorado, Montana, and Wyoming. Common in Arctic and alpine regions in Europe, Iceland, and Fennoscandia. August.

Notes A distinctive *Galerina* because of the membranous white ring that makes it look like a tiny *Pholiota*. *G. marginata* also has a white membranous ring but spores are verrucose. Potentially deadly.

- Cap hemispherical, smooth, greasy, orange-brown, not striate
- Gills cinnamon, separated
- Stem long, pale, darkening at base
- Ring white, membranous
- Spores almost smooth
- In moss

Reference Gulden 2010. CLC 1864.

Galerina vittiformis **f. *bispora*** A.H. Sm. & Singer

Cap 0.5–1 cm wide, bell-shaped, convex, red-brown, drying lighter from center outward to buff (hygrophanous), striate almost to center when moist, smooth. **Gills** deeply adnexed, a bit separated, orange-brown. **Stem** 2–4 × 0.1 cm, long and thin, equal, light golden-orange; whole stem pruinose (use hand lens). **Ring** absent. **Flesh** watery gray. **Odor** absent. **Spores** 10–11 × 6–7 µm, almond-shaped, verrucose; basidia 2-spored; cystidia ampul-shaped. No cystidia on cap.

Ecology and Distribution Reported from the Beartooth Plateau in Montana, in moss. The 2-spored variety is also found in the Arctic, specifically in Iceland and Norway. August.

Notes *Galerina atkinsoniana* is macroscopically and microscopically similar except that it has cystidia on the cap. Both species have 2-spored varieties. *G. alpestris* has primarily 4-spored basidia.

- Cap bell-shaped, orange to red-brown, striate
- Gills orange-brown
- Stem long and thin, pale, totally pruinose
- Spores verrucose; basidia 2-spored in ours
- In moss

References Armada et al. 2024; Gulden 2010. CLC 3602.

Hypholoma elongatum (Pers.) Ricken

Cap 1–3 cm wide, convex with small dome or indentation in center, dark caramel color, dark ocher, olive-yellow, more yellow-brown on margin, hygrophanous, striations showing on drying, smooth, greasy; margin turned down at first. **Gills** adnate or deeply adnexed, a bit separated, golden, orange-brown, olive-brown. **Stem** 2–4 × 0.2–0.5 cm, equal, bit undulating, pale ocher at apex, darkening toward base to olive-brown with copper hue, to olive-black, smooth, greasy. **Flesh** olive brown. **Odor** possibly of rubber. **Spores** purple-black, golden in KOH, 9–11 × 7–8 µm, broadly ellipsoid, germ pore not obvious, smooth; chrysocystidia present.

Ecology and Distribution Reported from the alpine of Montana and Wyoming, in mossy seeps. Also known from Arctic-alpine habitats with *Sphagnum* in Fennoscandia and the Alps, and at lower elevations in the Pacific Northwest. August.

Notes Recognized by the dark spore print, yellow olivaceous colors, and smooth cap and stem. *Hypholoma polytrichi* has smaller spores and *H. udum* has verrucose spores; both are found in similar habitats. Similar in aspect to smooth, greasy Hygrocybes.

- Cap convex, smooth, orange with olive hue
- Gills golden, orange-brown, olive-brown
- Stem pale at apex, darkening at base
- Spore print purple-black
- In mosses, mainly *Sphagnum*

Reference Knudsen and Vesterholt 2008. CLC 3901.

Loreleia marchantiae (Singer & Clémençon) Redhead, Moncalvo, Vilgalys & Lutzoni

Cap 0.5–1.5 cm wide, flat to slightly dished in center, orange, pale orange, hygrophanous, smooth; margin indistinctly striate. **Gills** decurrent, triangular, pale yellow-orange, only a few reach cap edge (10–14), well separated. **Stem** 2.5 × 0.1 cm, thin, undulating, pale orange, smooth or slightly pruinose at top. **Flesh** watery orange. **Odor** not noted. **Spores** white, 8.5–11 × 5–7 μm, smooth, ellipsoid, not amyloid.

Ecology and Distribution Rare, reported from the San Juan Mountains in Colorado at high elevations near a disturbed mine entrance; on the liverwort *Marchantia*. Known from Alaska, the Alps, Greenland, and Norway and also from sea level in the Netherlands. August.

Notes Although known as an Arctic-alpine fungus, this species can occur at lower elevations in pioneering, disturbed sites or on peaty soils. It is smaller with fewer gills than *L. postii,* which is more common on burns.

- Cap small, orange, dished in center
- Gills decurrent, whitish, few in number
- Stem thin, pale orange
- Spores white
- On the liverwort *Marchantia*

References Armada et al. 2024; Senn-Irlet et al. 1990. CLC 1419.

Loreleia postii (Fr.) Redhead, Moncalvo, Vilgalys & Lutzoni

Cap 0.5–1.5 cm wide, shallow convex with depressed center, and margin rolled down or under, bright burnt orange, smooth, greasy. **Gills** decurrent, well separated, bright white, cream, about 24 reach cap edge. **Stem** 1–1.5 × 0.1–0.2 cm, pale orange, pale yellow-orange, smooth, inserted in the liverwort *Marchantia*. **Flesh** white in stem. **Odor** not distinct. **Spores** white, 7–10 × 4.5–6.5 µm, ellipsoid, smooth, not amyloid.

Ecology and Distribution In the Rocky Mountains, more common in the subalpine, but occasional in alpine pioneering situations such as burned, disturbed, or peaty soils where the liverwort *Marchantia* is present. Also reported in Arctic areas of Norway and Alaska. July.

Notes The similar *L. marchantiae* is distinguished by a smaller cap, fewer gills, and slightly larger spores.

- Cap convex with depressed center, bright burnt orange
- Gills decurrent, bright white
- Stem pale orange
- Spores white
- On the liverwort *Marchantia* in pioneering habitats

Reference Senn-Irlet et al. 1990. CLC 3192.

Mycena citrinomarginata Gillet

Cap 0.8–2.0 cm wide, bell-shaped, then opening, olivaceous at first becoming bright yellow, striate almost to center, smooth. **Gills** adnate with tooth, whitish or with yellow tinge; margin yellow. **Stem** 2.5–5 cm x 0.1–0.2 cm, long and thin, pale yellow with olive hues. **Odor** not distinct. **Spores** white, 7–12 × 4–6 µm, ellipsoid, smooth, amyloid.

Ecology and Distribution In vegetation, often in open meadows in moss. Rare in the Rocky Mountains, reported from the Montana alpine. Known from Arctic-alpine habitats in Greenland, Svalbard, Iceland, Fennoscandia, and the Alps. July.

Notes Recognized by the yellow cap, yellow gill edges, and olivaceous colors when young. *M. olivaceomarginata* is similar and can't be ruled out. This is one of the few Mycenas reported from the Rocky Mountain alpine.

- Cap bell-shaped, bright yellow, striate
- Gills white with yellow edge
- Stem yellow to olivaceous
- Spores white
- In alpine or Arctic vegetation, including mosses

References Gulden and Jenssen 1988; Borgen 1993. CLC 1792; CLC 1917.

Mycena cf. *pasvikensis* Aronsen

Cap 1–2 cm wide × 0.6–1.2 cm high, bell-shaped, conic-convex, with rounded top, dark gray-brown to light gray-brown, darker in center, smooth, striate up to center, some with hoary coating; margin pleated. **Gills** sinuate, with a tooth, narrow, well separated, white to gray; edges concolorous. **Stem** 2–4 × 0.1–0.2 cm, long and thin, equal, curved, whitish on top and grayish below, at least partly pruinose. **Flesh** grayish, shiny-silky; stem rubbery. **Odor** of bleach, possibly farinaceous. **Spores** white, 9–12 × 5–6 µm, cylindrical to ellipsoid, smooth, amyloid; tissue pink in Melzers. Cystidia clavate with branches; cap skin of inflated cells.

Ecology and Distribution On willow debris in wet alpine areas of the Beartooth Plateau in the Rocky Mountains. Rare, so far only reported from alpine areas of Norway in association with *Salix*. July to August.

Notes This species likely has been misidentified as *M. cinerella* but appears to be more of an Arctic-alpine species with willow debris. It is tough-fleshed and not fragile. DNA sequencing is needed to confirm the tentative ID.

- Cap bell-shaped, gray, grayish brown, striate
- Gills sinuate, white to gray, separated
- Stem long and thin, grayish, tough
- Spores white, smooth, amyloid
- In *Salix* debris

Reference Aronsen and Læssøe 2016. CLC 3553.

Omphalina* cf. *rivulicola (J. Favre) Lamoure

Cap 1–3 cm across, deeply funnel-shaped, liver brown when fresh, drying pale brown to almost white, smooth, greasy; margin rolled under at first, wavy in age, slightly pleated. **Gills** well separated, running down the stalk, pale cream with edges that darken. **Stem** 2–5 × 0.1–0.4 cm, long and thin, smooth; pale brown, white at the base. **Flesh** watery cream. **Odor** absent. **Spores** white, 8–11 × 5–7 μm, ellipsoid, smooth.

Ecology and Distribution In the alpine, scattered in moss along streams and around ponds. Common, but missing in years of high fast runoff when streambanks are scoured. Known from the Colorado and Montana Rocky Mountains; also found in Arctic-alpine habitats in Iceland, Svalbard, Greenland, and the Alps. July to August.

Notes The funnel shape is reminiscent of a *Clitocybe*, but the cap is more delicate and pales on drying. Other funnel-shaped fungi in Arctic-alpine habitats, such as the Arrhenias, are smaller with shorter stems. Sequencing is necessary to distinguish it from the similar *Omphalina chionophila* Lamoure.

- Cap brown, pale brown, deeply funnel-shaped, delicate
- Gills decurrent, whitish cream
- Stem long and thin, pale brown
- Spores white
- In vegetation and moss along streams and ponds

References Armada et al. 2024; Knudsen and Vesterholt 2008. CLC 2295.

Rhizomarasmius epidryas (Kühner) A. Ronikier & Ronikier

Cap 5–10 mm across, convex with a slightly depressed center; variable in color, orange-brown, pale yellow-brown with darker center; margin pleated. **Gills** broadly attached or running slightly down the stalk, a bit thick, pale cream, sparse (only 15 to 20 gills). **Stem** 0.5–2 × 0.1–0.2 cm, equal, dark red-brown to black, minutely velvety over whole length (use a hand lens). **Flesh** thin, tough, and pliable in stem. **Odor** not distinctive. **Spores** white, 8–11 × 5–7 µm, almond-shaped; cystidia present.

Ecology and Distribution Apparently a decomposer found on dead parts of *Dryas* in Arctic and alpine habitats. In North America, reported from a few alpine locations in the Rocky Mountains, such as Loveland Pass, Canada, and Alaska; also known from AA habitats in Greenland, Scandinavia, Romania, Russia, and the Alps. August.

Notes The strict association with *Dryas* and the tough black velvety stems make this species easy to recognize. It does not appear to be parasitic on *Dryas*. Molecular data place it near genus *Marasmius*. It has a circumpolar distribution in cold climates and is an iconic Arctic-alpine fungus.

- Cap orange-brown, pleated, tough
- Gills widely separated, cream
- Stem long, thin, blackish, velvety, tough
- Spores white
- Attached to old parts of *Dryas*

Reference Ronikier and Ronikier 2010. CLC 3818.

Rickenella fibula (Bull.) Raithelh.

Cap 0.5–1.0 cm wide, convex with sunken center, dark orange, center darker, fading in age, faintly striate, smooth. **Gills** very decurrent, separated, sometimes forked, whitish, pale orange. **Stem** 1–3 × 0.1 cm, long and thin, pale orange, smooth but minutely pubescent, inserted in moss. **Flesh** pale, tough, rubbery. **Odor** not distinct. **Spores** white, 5.5–7.5 × 2–3 µm, ellipsoid to long cylindrical, smooth; cystidia on cap sparkle in the sun with a hand lens.

Ecology and Distribution Common in the Rocky Mountain alpine and subalpine habitats in mosses of wet areas along streams and around ponds. Also common in Arctic-alpine situations in Europe, Greenland, and Svalbard and the Pacific Northwest. August.

Notes Apparently a moss inhabitant that is extremely common in Arctic-alpine habitats, but when it pales, it can be overlooked. Its rubbery flesh can be confirmed with a "boink" test. Cystidia on the cap are an unusual feature. *Rickenella mellea* and *R. swartzii* also can be found in AA and other habitats; the former has a brownish cap and the latter a violaceous stem. Not related to other gilled mushrooms, but in the chanterelle family.

- Cap orange, with sunken center
- Gills pale orange, decurrent
- Stem pale orange, long and thin
- Flesh rubbery
- Spores white; cap with cystidia
- In moss in wet areas

References Korotkin et al. 2018; Knudsen and Vesterholt 2008. CLC 3827.

4

Mycorrhizal Mushrooms with Willow, Birch, and Dryas

Most mycorrhizal fungi live in forests and colonize tree roots. Therefore, it may be surprising to learn that these mutualistic fungi are prolific in lands above and beyond the trees. Miniature alpine and Arctic forests are composed of low willow (*Salix*) and bog birch (*Betula*), and sometimes alder (*Alnus*). *Dryas*, an alpine plant in the Rose family, is also mycorrhizal. These low woody plants form mutualisms with mycorrhizal fungi, just as trees do at lower elevations. Sugars (carbon) produced by the plants through photosynthesis leak from roots and are picked up by the fungi as nourishment. In return, the mycelium of the fungus accesses soil nutrients, particularly nitrogen in AA habitats, and shuttles it into plants, in the ectomycorrhizal association. There are other types of mycorrhizal mutualisms, such as those formed by AM fungi with forbs and grasses, and those composed of ericoid fungi and heaths and heathers, but these do not produce mushrooms. In the alpine and Arctic, ectomycorrhizal fungi can also occur with bistort and sedges such as *Kobresia*. The primary mycorrhizal fungal genera in AA habitats are some of the same ones found at lower elevations and in warmer habitats with trees, which include *Amanita, Lactarius, Russula, Cortinarius, Hebeloma, Laccaria, Leccinum, Entoloma* (some), and *Inocybe* s.l. However, species often differ from those at lower elevations.

How mycorrhizae function.

Mycorrhizal mushrooms are located near their plant hosts, but decomposer fungi also can take advantage of the microhabitat formed by willows, birch, and *Dryas*, and can appear nearby.

Mycorrhizal Plants of the Alpine and Arctic

Mycorrhizal host plants of the Alpine. Top row: A. Veined leaves of dwarf willow *Salix reticulata*. B. Pointed leaves of dwarf willow *Salix arctica*. Second row: C. Toothed leaves of shrub birch *Betula nana*. D. Bistort, *Persicaria viviparum* flowers. Bottom row: E. Scalloped leaves of mat plant *Dryas*. F. Flowers of *Dryas octopetala*, also called Mountain Avens.

Patchy distribution of mycorrhizal mat plants; *Dryas octopetala* mats in foreground and shrub and dwarf willow (*Salix*) in background.

Shrub willow *Salix planifolia* can be a few feet tall and has reddish branches.

Key to Mycorrhizal Genera

See pg 9 on how to determine spore color.

1. Fruiting body with cap, stem, and pores/tubes **Leccinum** (pg 196)
1. Fruiting bodies with cap, stem, and gills 2

PINK SPORES

2. Spore print pink, spores angular; gills pink (or white) **Entoloma** (pg 118)
2. Spores not as above, but gills can be white, pink, or pale salmon 3

WHITE TO YELLOW SPORES

3. Stem base with a cup (volva); spores white **Amanita** (pg 87)
3. Volva absent 4

4. Cap red, maroon, magenta, mauve, mottled red-brown, orange-brown (white for one); spores white or yellow with amyloid warts **Russula** (pg 199)
4. Cap without reddish tones, but may be lavender, purple-brown, or orange 5

5. Cap small, orange, often striate; gills orange or pink, rather thick; stem orange; spores white, spiny **Laccaria** (pg 178)
5. Not completely as above, but can have an orange cap and stem 6

6. Cap flat to sunken in center, yellow, orange, brown; gills yellow, salmon, orange, oozing milk or staining when cut; spores white with amyloid warts **Lactarius** (pg 186)
6. Not as above; spore print brown or rusty; many with a cobwebby veil 7

BROWN TO RUSTY SPORES

7. Caps in most species fibrous, woolly, or scaly (smooth and white in one); odor spermatic, fruity, spicy, or of burnt sugar 8

7. Caps mostly smooth, often greasy or viscid; odor of radish or indistinct 9

8. Cap flat to convex, ocher, brown, fibrous, wooly, or scaly; odor of burnt sugar; stem fibrous to scaly; spores brown, smooth **Mallocybe** (pg 168)

8. Cap pointed to convex, whitish, yellow-brown, brown, ocher, fibrous to scaly; stem fibrous or smooth; odor often spermatic; spores brown, smooth, or nodulose **Inocybe** (pg 142), **Inosperma,** and **Pseudosperma** (pg 168)

9. Cap small to medium, convex to flat, white, cream, brown, often two-toned, smooth, greasy; gills white to brown; odor often radish; spores brown, smooth to slightly rough **Hebeloma** (pg 128)

9. Cap small to large, pointed to convex, smooth, greasy or dry, yellow-brown, orange, brown, mauve, a few with lilac tints; odor radishy or not; spores brown, rusty red-brown, always with warts **Cortinarius** (pg 92)

Amanita in Alpine and Arctic Habitats

The genus *Amanita* is easily recognizable in Arctic and Alpine habitats as it has all the features of lower-elevation species, including a volva, free gills, sometimes white patches on the cap, and white spores; plus species mostly maintain their normal size. However, it can be disconcerting to see Amanitas fruiting in an open landscape and not under forest trees, since they are mycorrhizal fungi. The hosts are *Salix* species, either shrub willows a few feet tall, or dwarf willows a couple inches high. In the latter case, the mycorrhizal fungus towers over the miniature alpine *Salix* forests in a reversal of size.

The diversity of *Amanita* is limited in AA habitats, and distributions differ across North America. *Amanita nivalis* appears to be more common in the Southern Rocky Mountains. *Amanita groenlandica* is not reported from Colorado but is common on the Beartooth Plateau, and in Arctic Alaska and Canada. *Amanita arctica* is reported on the Beartooth Plateau and from Schefferville in eastern Canada.

Young *Amanita nivalis* with pink gills and saccate volva.

In addition to the three species reported here, *Amanita mortenii* has been found with birch in Greenland, and *Amanita muscaria* is occasionally reported in AA habitats but usually in the low alpine-subalpine zone near trees, and *Amanita regalis* occurs in Alaska.

Key to Alpine and Arctic Amanita

1. Cap brown to grayish brown, with patches of tissue; stem with "adder" pattern; volva saccate **Amanita groenlandica**
1. Cap white, cream, pale gray, gray, with patches of tissue or not; stem floccose; volva saccate or appressed 2

2. Cap grayish when young, paling to whitish in age, small, delicate; volva saccate **Amanita nivalis**
2. Cap whitish to cream, tall and slender; volva appressed **Amanita arctica**

Amanita cf. *arctica* Bas, Knudsen & T. Borgen

Cap 4–7 cm across, convex to flat, pure white turning cream, with scattered tissue patches or none; margin pleated. **Gills** free, pure white to cream; edges floccose. **Ring** absent. **Stem** 5–7 × 1–1.5 cm wide, enlarging toward base, white, with fine pattern. **Volva** appressed to stem, flaring at top, white with gray cast inside. **Odor** unpleasant in age. **Flesh** white. **Spores** white, 12–14 × 10–12 μm, almost round, not amyloid.

Ecology and habitat Mycorrhizal with willows, occurring in the low alpine with shrub willow *S. planifolia* on the Beartooth Plateau. Known from Greenland, eastern Arctic Canada, and Finland. Rare. August.

Notes Young specimens are pure white and turn dark cream, without gray as for *A. nivalis*. *Amanita groenlandica* has a buff to brownish cap. There is some taxonomic confusion with *A. oreina* that needs to be sorted out.

- Cap white, then cream with patches
- Gills free, white; stem white to cream
- Volva flaring at top
- Spores white
- With willows

References Armada et al. 2034; Knudsen and Borgen 1987; Hutchison et al. 1988. CLC 3412.

Amanita groenlandica f. *alpina* C.L. Cripps & E. Horak

Cap 4–8(12) cm across, robust, convex, pale salmon buff to brown in age with patches of universal veil; margin slightly striate. **Gills** free, crowded, white to pale orange. **Stem** 4–10 × 1–3 cm, enlarging toward base, whitish, covered with "adder" pattern of floccules. **Volva** sac-like, whitish, fragile. **Flesh** white. **Odor** fruity or unpleasant in age. **Spores** white, 10–13 × 10–12 um, smooth, almost round, not amyloid.

Ecology and Habitat Mycorrhizal with dwarf willows *Salix reticulata* and *S. arctica* and shrub willow *S. glauca*. Common on the Beartooth Plateau in Wyoming and Montana. Also in Greenland, Alaska, and Canada. July and August.

Notes Recognized by the salmon to brown cap, lack of a ring, and robust stature. Freezing or drying can alter the color to brownish gray. *A. nivalis* is smaller, more delicate, light gray to white, and *A. arctica* is white to cream.

- Cap salmon-brown with patches of tissue
- Gills free, white; stem white to cream, with adder pattern
- Volva saccate
- Spores white
- With willows

References Cripps and Horak 2010. Hutchinson et al. 1988. CLC 2328.

Amanita nivalis Grev.

Cap 3–7 cm across, convex to flat, gray becoming white in age, occasionally with one volval patch, smooth, slightly greasy; margin striate. **Gills** free, white, pink when young; edges floccose. **Stem** 4–6 × 0.5–1.5 cm, enlarging toward base, white, minutely floccose the whole length. **Ring** absent. **Volva** cup-shaped, flaring, lobed, white. **Flesh** white. **Odor** not distinct. **Spores** white, 10–12 × 8–11 μm, subglobose, smooth, inamyloid.

Ecology and Habitat In North America, known from the Colorado alpine, often with dwarf willow *Salix reticulata*. Rare in Montana. Common with willows in Arctic-alpine habitats in the Northern Hemisphere. July and August.

Notes The famous "ringless, snow Amanita" was first described from the Scottish Highlands and occurs in Iceland, Norway, Sweden, and Finland with willows. It looks like a miniature *A. vaginata* but has pink gills when young, and the cap pales or turns white in age.

- Cap small, gray to white, striate
- Gills free, white/pink; white stem, minutely floccose
- Volva flaring
- Spores white
- With willows

Reference Cripps and Horak 2010. DBG 21387 (Evenson).

Cortinarius in Alpine and Arctic Habitats

A diversity of *Cortinarius* species thrives in alpine and Arctic habitats, from large fleshy types to little brown kinds that can all look similar. Like their counterparts in subalpine and boreal habitats, *Cortinarius* is recognized by the presence of a cobwebby cortina that connects the cap margin to the stem in young specimens (look closely), and rusty reddish-brown spores—although brown spores are also possible. The odor is mostly raphanoid, which means "radishy." Microscopically, the genus has almond-shaped, ellipsoid, or almost round spores, and spores are always warted. Cystidia are mostly absent.

A helpful way to think of *Cortinarius* s.l. diversity is in terms of historical subgroups. *Myxacium* species have slimy caps and stems and are particularly abundant, although not diverse, in AA habitats. Those in the *Anomali* group have convex, pale caps, often lavender tints on the gills or stem, and rather round spores. The tiny, tedious Telemonias are everywhere in cold northern climes. Most have smooth, silky, pointed caps and silky white bands on their thin stems. Many are a deep umber brown or orange color. The *Dermocybe* group is distinguished by brightly colored yellow, orange, or red gills, and several occur in AA habitats. All Cortinarii are mycorrhizal and occur near willows, birch, or *Dryas*.

Key to Alpine Cortinarius

MYXACIUM

1. Cap and stem slimy; fruiting body slippery 2
1. Not as above 4

2. Cap yellow; stem white, clavate **Cortinarius vibratilis**
2. Cap pale to dark orange-brown, often darker in center; stem white, tapered 3

3. Cap 1–4 cm; stem white, base fleshy; with dwarf willows **Cortinarius alpinus**
3. Cap larger, 3–9 cm; stem white, base woody; with shrub willows **Cortinarius absarokensis**

DERMOCYBE

4. Gills bright yellow or bright coppery orange when young 5
4. Gills not as above 9

5. Gills coppery orange; cap coppery; stem with red fibrils **Cortinarius uliginosus**
5. Gills bright yellow, then orange-brown; stem not as above 6

6. Medium types; stem typically longer than cap width 7
6. Small, short types; stem typically shorter than cap width 8

7. Cap more yellow-brown, brown, convex **Cortinarius cinnamomeoluteus**
7. Cap more orange-brown, red-brown, tall campanulate when young **Cortinarius ferruginosus**

8. Cap hemispherical, brown, rough **Cortinarius sp.**
8. Cap convex with a pointed apex, orange-brown, smooth **Cortinarius pratensis**

ANOMALI GROUP

9. Cap pale to medium brown; gills lilac, buff, gray; spores roundish 10

9. Not as above; spores almond-shaped to ellipsoid 12

10. Cap large (3–7 cm), brown; gills lilac-gray, mauve; in krummholz **Cortinarius 'caninoalpinus'**

10. Cap smaller; gills lilac or not; often cespitose; clearly with willows 11

11. Gills often lilac; cap pale orange-brown, minutely textured/areolate **Cortinarius albidipes**

11. Gills typically grayish brown; cap pale brown, smooth **Cortinarius tabularis**

TELEMONIA

12. Cap relatively large (2.5–6 cm) *and* dark brown; stem 0.5–2 cm wide 13

12. Not as above; stem slimmer 14

13. Cap umber brown, hoary at first; gills mauve; often cespitose **Cortinarius saturninus**

13. Cap center dark brown, not hoary; gills pale brown **Cortinarius piceidisjungendus**

14 Stem long and slim 2–3 mm wide, apex violet; cap brown **Cortinarius pulchripes**

14. Not as above 15

15. Cap, gills, stem orange; stem with white bands of tissue; odor pungent **Cortinarius parvannulatus** (= *C. cedriolens*)

15. Not as above 16

16. Cap light brown, brown; gills pale brown, pale orange-brown 17

16. Cap dark brown, red-brown; gills orange-brown, red-brown 18

Cortinarius saturninus in cluster. CLC 2312.

17. Cap pale brown, with white hair tufts when young; gills pale **Cortinarius expallens**
17. Cap medium brown, smooth, silky; gills pale orange **Cortinarius fuscoflexipes**

18. Cap red-brown; gills red-brown; stem with white tissue **Cortinarius hinnuleus var. favreanus**
18. Cap dark brown, gills orange-brown; stem covered with white fibrils **Cortinarius umbilicatus**

Additional species reported from the North American alpine by M. M. Moser (1993)

C. albonigrellus J. Favre

C. atroalbus M.M. Moser

C. chrysomallus Lamoure

C. ferrugineifolius M.M. Moser = *C. subrigidipes* = *C. paraphaeo-chrous* M.M. Moser (*C. pangloius* M.M. Moser) *C. galerinoides* Lamoure

C. hybospermus M.M. Moser

C. inops J. Favre

C. laetus M.M. Moser

C. minutalis Lamoure

C. pauperculus J. Favre

C. phaeochrous J. Favre

C. phaeopygmaeus J. Favre

C. pusillus M.M. Moser, McKnight, Sigl

C. saniosus Fr. = *C. rufoanuliferus* M.M. Moser & McKnight

C. stenospermus Lamoure

C. tenebricus J. Favre

C. uraceus Fr.

C. vulpicolor M.M. Moser & McKnight

Cortinarius ferruginosus with mature, opened caps.

Cortinarius absarokensis M.M. Moser & McKnight

Cap 3–7(–9) cm, robust, fleshy, convex, often domed, uplifted in age, cream-buff, pale orange, caramel color, orange-brown, smooth, glutinous. **Gills** notched, well separated, becoming broad, cream, pale brown, pale orange, orange-brown. **Stem** 4–8(–10) × 1–2.5 cm, equal to tapered, white, glutinous, smooth above a glutinous **cortina**, fibrous below, woody and darker at base. **Flesh** white, golden brown; woody in base. **Odor** not distinct. **Spores** rusty, 13–15 × 7–9 µm, almond-shaped, verrucose.

Ecology and Distribution Primarily known from the Rocky Mountain alpine and subalpine from Colorado to Montana, where it is mycorrhizal with shrub willows *S. planifolia* or *S. glauca*. Reported a few times from the Alps. July to August.

Notes Named for the Absaroka Mountains in Montana-Wyoming by Austrian mycologist Meinhard Moser; this species is a larger version of *Cortinarius alpinus* with a woody base and association with shrub willows. Similar to the subalpine *Cortinarius trivialis*.

- Cap robust, pale buff to orange-brown, slimy
- Gills pale, becoming orange-brown
- Stem white, slimy, base woody; cortina slimy
- With shrub willows

Reference Moser and McKnight 1987. Several sequenced.

Cortinarius albidipes Peck

Cap 2–5(–8) cm, hemispherical, convex, or almost flat, thick-fleshed, pale yellow-brown, pale brown, pale orange-brown, hoary at first, appearing smooth but finely textured (use hand lens), becoming areolate in age; margin inrolled at first, often pale lilac. **Cortina** grayish-lilac fading to whitish, often copious. **Gills** narrowly attached, almost subdecurrent, narrow to broad, grayish-lilac, pale milk coffee. **Stem** stout, short or not, 1.5–7 × 0.5–3 cm, clavate, tapered when clustered, smooth to fibrous, dingy white, lilac at apex, with faint pale yellow-orange bands. **Flesh** white, grayish lilac in stem apex, solid, firm. **Odor** none or faintly raphanoid. **Spores** brown, 8.5–9.7 × 6.3–7.3 μm, round, verrucose.

Ecology and Distribution Reported here as mycorrhizal with willows in the Rocky Mountain alpine (Montana, Wyoming, Colorado). More typically known from temperate, boreal and subalpine forests across North America with many broadleaf trees; often cespitose. August.

Notes Recognized by the pale cap that can be areolate, lilac gills, and sometimes squat stature (more squat in the alpine). *Cortinarius tabularis* usually lacks lilac colors, has a smooth cap and smaller spores. First described by Peck from the eastern United States.

- Cap pale brown, finely textured to areolate, fleshy
- Gills lilac; cortina grayish lilac
- Stem white with lilac apex; often cespitose
- With willows, perhaps *Dryas*
- Spores round, verrucose

Reference Dima et al. 2021. CLC 3539, sequenced.

Cortinarius alpinus Boud.

Cap 1–4 cm, conic-convex, convex, caramel, orange-brown, darker in center, sticky, glutinous; margin turned down. **Gills** attached, pale cream, pale orange, foxy brown, sometimes with lavender hue. **Stem** 2–6 × 0.5–1 cm, equal or tapered to a point; white, glutinous, slippery; smooth above a glutinous ring where rusty spores stick, fibrous below. **Flesh** white, fleshy in stem base. **Odor** absent. **Spores** rusty, 11–14 × 7–9 µm, ellipoid to slightly almond-shaped, verrucose.

Ecology and Distribution Common in the Rocky Mountain alpine; one of the first mushrooms to appear in late July. Mycorrhizal with dwarf willows. Well known from many Arctic and alpine habitats around the Northern Hemisphere, including Greenland, Canada, Iceland, the Alps, and Fennoscandia. July to August.

Notes Also called *C. favrei,* after the father of alpine mycology, Jules Favre, who collected in the Swiss Alps. *Cortinarius absarokensis* is larger with a woody stem and occurs with shrub willows.

- Cap light to dark orange, slimy
- Gills pale to dark orange; ring glutinous
- Stem white, fleshy, slimy, tapered
- With dwarf willows

Reference Moser 1993. CLC 3879, sequenced.

Cortinarius 'caninoalpinus' ad. int. (C.L. Cripps & Peintner)

Cap 3–5(–7) cm, hemispherical, broadly convex, slightly domed, light brown with a hoary coating, becoming foxy orange-brown, smooth, cracking in age, dry, dull; margin turned down or under. **Cortina** whitish, soon gone, cobwebby or a bit membranous. **Gills** notched, becoming broad, lilac gray, then lilac-brown, grayish brown, orange-brown. **Stem** sturdy, 4–6(–8) × 0.7–1.5 cm, enlarging toward slightly swollen base, grayish lilac at apex, whitish below, with sparse brown bands, silky above ring, fibrous below; grayish mycelium at base. **Flesh** firm, gray at top of stipe. **Odor** indistinct, or faintly fungoid, raphanoid. **Spores** brown, 8.0–10 × 6.0–7.5 μm, subglobose, subellipsoid, strongly verrucose.

Ecology and Distribution In the low alpine or krummholz zone, near dwarf willow *Salix reticulata, Dryas octopetala*, and shrub *Salix,* but conifers (spruce) near most collections. So far only known from the Rocky Mountains. August.

Notes Molecularly closely related to *C. caninus,* which is a forest species in North America and Europe. *Cortinarius albidipes* also has lilac gills but is usually squatter, and the cap is rougher. *Cortinarius tabularis* usually lacks lavender tints. All have roundish spores.

- Cap broadly convex, smooth, foxy orange-brown
- Gills lavender at first; cortina whitish
- Stem white; spores roundish, warted
- In krummholz zone near willow, *Dryas*, and conifers

Reference Dima et al. 2021. CLC 3288, sequenced.

Cortinarius cf. *cinnamomeoluteus* P.D. Orton

Cap 2–4.5 cm, conic-convex, broadly convex, or almost flat with small umbo, medium to dark yellow-brown in center, golden toward the margin, indistinctly radially fibrous to silky smooth; margin turned in at first, uplifted in age. **Cortina** yellow, cobwebby, ephemeral, often not apparent. **Gills** adnexed or toothed, bright yellow, golden, broad in age. **Stem** 2.5–4 × 0.3–0.6 cm, mostly equal, pale yellow, yellow, old gold, fibrous or silky. **Flesh** pale or dark golden. **Odor** absent, or faintly raphanoid. **Spores** ocher, 9–10(–15) × 5–6 μm, ellipsoid to almond-shaped, verrucose.

Ecology and Distribution Reported from subalpine and alpine habitats with willow, alder, and possibly birch in Europe and North America; especially in marshy and Arctic-alpine habitats. August in the alpine.

Notes The bright yellow gills identify this as a *Dermocybe*. It appears to be quite common in AA habitats and is well known in Scandinavian countries. It is larger than *C. polaris* or *C. pratensis,* which also have yellow gills. The cap of *C. ferruginosus* is more reddish brown and campanulate.

- Cap with yellow-brown center, golden yellow at margin
- Gills bright yellow to golden; cortina yellow
- Stem pale to dark yellow or golden
- With willow, alder, or birch

References Huymann et al. 2024; Knudsen and Vesterholt 2008. CLC 2860 sequenced. CLC 3558 (above).

Cortinarius expallens M.M. Moser

Cap 1–3 cm across, sharply conic-convex, or almost flat with small nipple, brown, pale grayish brown, reddish brown, covered with fine tufts of white hairs when young. **Gills** adnate to sub-decurrent, well separated, pale orange, milk coffee. **Cortina** white. **Stem** 3–5 × 0.3–0.5 cm, undulating, cream, watery brown where touched, with peronate white sock that flares up as a ring. **Flesh** cream, brown in stem. **Odor** absent. **Spores** ocher, 8–10 × 5–6 µm, ellipsoid, verrucose.

Ecology and Distribution We report this species on the Beartooth Plateau with willows, as did M.M. Moser and McKnight. Also known from the Alps and Northern Europe from subalpine and alpine habitats with birch and willow. August.

Notes The related subalpine *C. hemitrichus* is known as the frosty webcap because of the white tufts on the cap. Favre (1955) described the alpine forms, *C. mucronatus* and *C. hemitrichus f. improcerus,* which are the same, and have illegitimate names; our fungus matches Favre's type, and the appropriate name is *C. expallens.*

- Cap pale brown, pointed, with white tufts
- Gills pale orange, milk coffee, well separated
- Stem white, with sock-like ring
- With willows or birch

References Armada et al. 2024; Moser and McKnight 1987; Favre 1955. CLC 3590, Genbank 192584.

Cortinarius ferruginosus (A.H. Sm.) Ammirati et al.

Pileus 1.5–3.5(–5) cm across, young tall conic-convex with rounded umbo or strongly campanulate when young, convex or applanate, even dished in age, with wide rounded umbo, orange-brown, red-brown, sometimes with an olive hue, golden at margin, dull to shiny, fibrous; margin turned in at first. **Gills** adnexed, well separated, golden yellow, orange-brown. **Cortina** golden yellow. **Stem** 2–5.5 × 0.3–0.8 cm, equal or enlarged toward base, pale golden at apex, golden, olivaceous brown toward base, some pink or wine color at base, with brown fibrils. **Flesh** golden yellow, brownish, occasionally olive in the base. **Odor** indistinct, fungoid, or pungent. **Spores** rusty, 8–10 × 5–7 µm, almond-shaped, verrucose.

Ecology and Distribution In boggy, mossy areas of the Beartooth Plateau in Montana and Wyoming at alpine elevations with willows. Also reported from lower elevations almost to sea level in Washington state. August.

Notes This large, red-brown *Dermocybe* is found in mossy habitats with willows. Smith (1939) published this species as an *Inocybe* likely because of the unusual shape of young caps, but this has been corrected (Niskanen 2014).

- Cap tall campanulate when young, then convex
- Cap golden/orange/red-brown with yellow margin
- Gills orange; stem golden
- With willows in wet areas

References Smith 1939; Niskanen 2014. CLC 3611, sequenced.

Cortinarius fuscoflexipes M.M. Moser & McKnight

Cap 2–3.5 cm across, conic-convex with nipple-like umbo, brown, umber brown, blackish in center, some coppery tones, hygrophanous, radially fibrous, silky-shiny; margin whitish, turned down. **Cortina** buff, soon gone, fibrils left on stem. **Gills** sinuate, soft orange, pale orange-brown, brown, well separated. **Stem** long, slim, flexuous, 3–7 × 0.3–0.4 cm, brownish, covered with white fibrils, can be bluish at apex; white mycelioid at base. **Odor** not distinctive. **Flesh** brown. **Spores** 8–10 × 5.5–6 μm, ellipsoid to slightly almond-shaped, verrucose.

Ecology and Distribution Cespitose, under shrub willow *Salix planifolia* and spruce on the Beartooth Plateau and in Yellowstone National Park. August.

Notes First described from Mount Washburn with *Salix* and spruce by M.M. Moser, not far from our sighting. The bluish stem base was not originally noted in ours. The similar *Cortinarius dryadophilus* is recorded from the Alps; it has a fruity odor.

- Cap umber brown with tiny blackish umbo; silky-shiny
- Gills orange to brown, separated
- Stem covered with white fibrils
- Cespitose under willows with spruce nearby

References Armada et al. 2024; Moser and McKnight 1987. CLC 3556, Genbank PQ167990.

Cortinarius hinnuleus var. *favreanus* Bon

Cap 1.5–3.0 cm in diameter, shallow conic-convex to almost applanate, some with sunken center or slight umbo, date brown, red-brown, smooth, greasy, radially streaked in age; margin orange-brown. **Gills** deeply marginate, well separated, dark orange-brown, red-brown, rusty-brown. **Cortina** whitish. **Stem** 2–4 × 0.4–0.6 cm, equal or tapered at base, watery orange-brown, covered with whitish fibrils, forming bands. **Flesh** dark watery orange. **Odor** earthy. **Spores** rusty, ellipsoid, verrucose, 8.5–10 × 5.6 µm.

Ecology and Distribution Occurring in the alpine with *Salix* species; reported from the San Juan Mountains of Colorado and Two Ocean Mountain in Wyoming, where it is apparently common with dwarf willow *S. arctica*. First reported from the Alps but not well known. Late July to August.

Notes The dark red-brown, smooth cap and rusty brown, thick, and well-spaced gills distinguish this taxon. Moser first reported the form *graveolens* with an earthy odor, which fits ours. *Cortinarius hinnuleus* proper lacks the reddish tones in the gills. The taxonomy of this group needs work.

- Cap dark red-brown, smooth
- Gills orange-brown to red-brown, thick, well spaced
- Odor earthy, sometimes strong
- With willows

References Jamoni 2008; Moser 1993. CLC 1820, sequenced.

Cortinarius parvannulatus Kühner

Cap 0.5–1.5 cm across, sharply conic-convex, some with pointed umbo, dark orange-brown, ocher on drying, smooth, greasy, shiny-silky, some slightly striate or streaky; margin with white tissue. **Gills** adnexed, thickish, well spaced, pale orange to orange-brown. **Cortina** whitish, typically as sub-membranous ring. **Stem** 1–2 × 0.2–0.3 cm, equal pale orange to ocher, rather smooth, with whitish tissue, occasional violet above ring, darker at base. **Flesh** orange. **Odor** pungent. **Spores** ocher, 8–9 × 5–6 μm, ellipsoid, verrucose.

Ecology and Distribution We report this species from the Beartooth Plateau, Montana, and Wyoming with dwarf and shrub willows. Known from the Alps and Northern Europe in alpine habitats with willows. August.

Notes Moser and McKnight report this mushroom from the Rockies as *Cortinarius cedriolens,* which is a synomym. The overall orange color and white veil tissue on the stem make it recognizable.

- Cap conic-convex, often with sharp umbo
- Cap, gills, and stem some shade of orange
- Stem with a whitish peronate sheath
- Odor of cedar, or pungent
- With shrub and dwarf willows

References Armada et al. 2024; Knudsen and Vesterholt 2008; Moser and McKnight 1987. CLC 3776, CLC 3114 (inset) sequenced.

Cortinarius* cf. *piceidisjungendus Kytöv., Liimat., Niskanen & Ammirati

Cap 2.5–5.0 cm across, robust, convex, slightly domed, umber brown, red-brown, lighter at margin, shiny-silky. **Gills** adnexed, well spaced, broad, milk coffee, darker in age, edges can be crenate. **Cortina** whitish buff, cobwebby, copious on cap margins of young specimens, remaining on stem. **Stem** stout, 2.5–4.5 × 0.8–1.5 cm, equal to clavate, dingy white, fibrous, dry. **Flesh** buff. **Odor** absent, or faintly raphanoid. **Spores** brown, 10–12 × 6.5–7.5 µm, ellipsoid, verrucose.

Ecology and Distribution Known from conifer forests in Washington, Alaska, Finland, and Sweden, but reported here with pure shrub willows *Salix planifolia* and *S. glauca* in late August in alpine habitat on the Beartooth Plateau.

Notes One of the larger *Cortinarius* species from the Beartooth Plateau, but smaller than is reported for subalpine forests. In the alpine, it is reminiscent of *C. absarokensis,* but the cap and stem are not sticky. Named for its habitat with coniferous trees, which were absent for these alpine specimens with willow.

- Cap dark brown, red-brown with lighter margin, silky, robust
- Gills well separated, pale brown to darker; stem white
- Copious white cortina when young
- With willows

Reference Liimatainen et al. 2015. CLC 3598 Genbank PQ167991 (almost matches holotype).

Cortinarius pratensis (Bon & Gaugué) Høil.

Cap 2–4 cm, hemispherical, convex, with low obtuse umbo, in age with sharp conical papilla, pale to dark chestnut brown, orange-brown, radially fibrillose to smooth, dry; margin clean. **Cortina** not noted. **Gills** adnexed, dark golden yellow, orange, becoming rust brown with ocher tint. **Stem** up to 4.5 × 0.6 cm, equal, yellowish becoming paler, fibrillose, dry, hollow. **Flesh** yellow in cap, turning olive green on exposure in base. **Odor** not distinct. **Spores** 7–9 × 5.5–6.5 µm, ellipsoid to almond-shaped, verrucose.

Ecology and Distribution In the Rocky Mountains with dwarf willow *S. reticulata* in alpine grasslands in August at 12,200 ft. in Colorado. Reported in dunes, grasslands, and heathlands in Norway, and possibly with willows. August.

Notes Our collection molecularly matches a Hungarian collection of this species. A fibrillose to smooth cap with a sharp umbo distinguishes it from other small alpine Dermocybes, such as *C. polaris* or *C. cinnamomeoluteus*.

- Cap chestnut brown, with dome or sharp papilla
- Gills golden or orange
- Stem pale yellow, fibrillose
- With dwarf willow *Salix reticulata* in open fields

Reference Høiland 1984. CLC 1438 (=ZT 9030) sequenced.

Cortinarius cf. *pulchripes* J. Favre

Cap 1–3 cm, conic-convex, then broadly convex to flat with acute umbo, medium to dark brown, hygrophanous, drying pale brown, center smooth, outward radially fibrous; margin whitish. **Gills** emarginate, pale medium brown, orange-brown with lilac tint. **Cortina** whitish. **Stem** 2.5–5 × 0.1–0.3 cm, long and thin, equal, undulating, watery brown, lilac at apex, covered with longitudinal white silky fibrils, which are soon gone. **Odor** weak. **Flesh** brownish mauve. **Spores** red-brown, 8–10.5 × 5–7 μm, ellipsoid, weakly verrucose.

Ecology and Distribution In the alpine with shrub willow *Salix glauca*, in the San Juan Mountains and on Loveland Pass in Colorado. Known from Europe and Fennoscandia with *Salix, Betula*, and *Alnus,* and from alpine France with dwarf willows. August.

Notes This tiny, slim *Telemonia* has a brown cap, orangish gills and a long, thin stem with whitish covering, tinted lilac at the apex. It is rarely recognized or rare. Our identification needs molecular confirmation.

- Cap conic, with umbo, umber brown
- Gills orange-brown with lilac tint
- Stem brown with whitish covering, and lilac apex
- Flesh mauve
- With willows

References Armada et al. 2024; Favre 1948; Breitenbach and Kränzlin 2000. JV32156F above (Norway); CLC 1683 inset; ZT 8081, sequenced.

Cortinarius saturninus (Fr.) Fr.

Cap 3–6 cm across, robust, conic-convex, convex, domed with distinct boss or not, brown, date brown, with grayish lavender tints, at first with hoary coating, smooth, satiny, cracking in age; margin lighter, turned down. **Cortina** pale whitish lavender, slightly membranous. **Gills** adnexed, well spaced, becoming broad, lavender, lavender-brown, orange-brown. **Stem** 3–7 × 0.5–2 cm, stout, clavate, slimming in age, lavender at apex, cream below, hoary, some with peronate sheath. **Flesh** lavender in top half; solid in stem. **Odor** mild to pungent. **Spores** 9–11 × 4–5 µm, narrowly almond-shaped to ellipsoid, verrucose.

Ecology and Distribution On the Beartooth Plateau, in dense clusters in alpine and Arctic habitats with shrubby willow species. Also known from Arctic Svalbard, France, Sweden, and Finland, where the hosts are willows or *Dryas*. August.

Notes A morphologically variable *Telemonia*. Caps are variable in size, and coloration and can be hoary or not. Thought to occur with other trees in the subalpine. Single, young specimens better fit the original description of *C. subtorvus*, which appears to be the same species.

- Cap dark brown, hoary or not, robust; often in clusters
- Gills lavender at first; stem white
- Flesh lavender when young
- With willows

Reference Liimatainen et al. 2017. CLC 2312, Genbank 168000. Also CLC 3095, CLC 3593.

Cortinarius sp.

Cap 1–2.5 cm, small, hemispherical, convex, or slightly conic-convex, brown, red-brown with golden or coppery orange tones, slightly textured, indistinctly appressed-squamose; margin turned down. **Cortina** of yellowish fibrils, soon gone. **Gills** adnexed, a bit separated, golden; edges fimbriate, lighter or not. **Stem** 2–3 cm × 0.4–0.6 cm, equal or tapered, whitish ocher, longitudinally fibrous with a few fibrils. **Flesh** watery yellow-brown. **Odor** not distinct, possibly earthy. **Spores** yellow-brown, 9–10 × 5.5–6.5 µm, ellipsoid to almond-shaped, verrucose.

Ecology and Distribution With dwarf willow *Salix arctica* in the Rockies, and possibly also *Dryas*. A possible endemic, so far reported only from Montana and Colorado. August.

Notes Other alpine Dermocybes are larger or have a smooth cap. At first considered to be *Cortinarius polaris,* which also has a small stature, but molecular data clearly separates the two. *Cortinarius polaris* is reported from alpine and Arctic habitats in Norway, Svalbard, the Alps, and Alaska.

- Small red-brown, minutely roughened cap
- Gills golden; stem whitish yellow
- With dwarf willows, especially *S. arctica*

Reference Høiland K. 1984. CLC 3219, CLC 2773 sequenced.

Cortinarius tabularis (Fr.) Fr.

Cap 1.5–6.0 cm, hemispherical, convex, pale brown, pale yellowish brown, sometimes darker red-brown in center and paler at margin, with hoary coating, marbled, smooth, shiny when dry; margin turned under or down. **Cortina** whitish buff, scant. **Gills** deeply sinuate, pale grayish white at first, then pale brown, pale orange-brown (rarely grayish blue). **Stem** 2.5–7(–9) × 0.4–1.0 cm, mostly equal, whitish, silky above, fibrous below, with faint sparse yellowish orange bands. **Flesh** whitish, gray at apex, yellowish at base. **Odor** raphanoid. **Spores** brown, 7–10 × 5–8 µm, subglobose to ellipsoid, verrucose.

Ecology and Distribution Confirmed from high elevations in the Rocky Mountains and Alaska in North America with willows and birch; and also Quebec. In Europe, known from birch or oak stands, and also Arctic-alpine habitats with *Salix* and *Betula*. Often cespitose. August.

Notes Collections from Arctic and alpine habitats reported as *C. anomalus* are more likely *C. tabularis*, which occurs with willows and birch. Unlike "*C. aninoalpinus*" and *C. albidipes*, the gills usually lack a lilac color; however, see the nanoform of *C. tabularis* on next page.

- Cap pale brown, hoary, shiny when dry
- Gills pale brown, not lilac, but may be grayish blue
- Stem equal, white, with faint yellow-orange bands
- Spores round, verrucose

Reference Dima et al. 2021. CLC 2131, sequenced.

Cortinarius tabularis (Fr.) Fr. (nanoform)

Cap 2.0–4.5 cm, hemispherical to convex, pale to medium yellow-brown, pale to medium brown, innately marbled, sometimes micaceous-shiny when dry; margin strongly turned in at first. **Cortina** perhaps whitish, but leaving orange-yellow fibrils on stipe. **Gills** adnexed, slightly separated, grayish (some with slight gray-lilac tones), then cream, remaining pale. **Stem** 2.5–3.5(–7) × 0.3–1.0 cm, clavate, narrow at apex, some with slight onion bulb and then pointed at base, smooth, whitish, cream, with sparse yellow-brown fibrils lower down; occasionally with grayish mycelium at base. **Flesh** watery gray or watery yellow-brown. **Odor** none or faintly raphanoid. **Spores** 6–9 × 5–7 µm, subglobose to ellipsoid, verrucose.

Ecology and Distribution Small forms of this species occur in Arctic and alpine habitats, primarily with dwarf willows such as *S. reticulata*. August.

Notes Our nanoforms are molecularly identical to the larger forms and are confirmed as the same species. However, with the small cap, gills verging on lilac, and clavate stem, these nanoforms could be misconstrued as other alpine species such as some forms of *C. epsomiensis* or *C. alpicola*.

- Cap small, hemispherical, pale brown, marbled, shiny
- Gills well separated, pale grayish, grayish lilac
- Stem short, clavate, white, with faint yellow bands
- Spores round, verrucose

Reference Dima et al. 2021. CLC 2320, sequenced.

Cortinarius uliginosus var. *uliginosus* Berk.

Cap 2–4 cm across, broadly convex with low umbo, medium brown with a strong orange cast, orange-brown, coppery, more yellow at margin, in center slightly bumpy, otherwise silky smooth, fibrous. **Gills** adnexed, broad or not, yellow at first, then bright orange, orange-brown. **Cortina** likely orange. **Stem** 3–4.5 × 0. 4–0.5 cm, equal, silky smooth, golden ground color with strong red-orange overlay, red-orange bands go down the stem. **Flesh** dingy cream pale yellow, olive at base. **Odor** not distinct. **Spores** rusty, 8.9 × 5.6 µm, almond-shaped, verrucose.

Ecology and Distribution With willow species in alpine habitats; reported from the San Juan Mountains and Niwot Ridge in Colorado. The form *uliginosus* is apparently restricted to willows. Also reported from Greenland in AA habitats and Europe in general. August.

Notes This *Dermocybe* is easily identified by its bright orange coppery colors. *Cortinarius uliginosus* proper is widespread in forests of Europe and North America, but this particular form is usually with willows.

- Cap orange, orange-brown, coppery, with yellow margin
- Gills orange
- Stem orange with coppery tones, red hairs
- With willows

Reference Knudsen and Vesterholt 2008. CLC 2277, Genbank PQ186727.

Cortinarius* cf. *umbilicatus P. Karst.

Cap 1.5–3 cm, broadly conic-convex, center can be depressed, dark brown, umber brown, outer part drying yellow-brown, greasy, smooth, with white margin. **Gills** adnexed, somewhat broad, well spaced, bright to dull orange. **Cortina** whitish. **Stem** 2–4 × 0.3–0.6 cm, equal, dark brown overlain with white coating or white bands. **Flesh** dark brown. **Odor** slightly fragrant or earthy. **Spores** 7.5–9 × 4.5 × 5.5 µm, ellipsoid, punctate.

Ecology and Distribution In alpine habitats with dwarf willows *S. reticulata* and *S. arctica*; reported in the San Juan Mountains of Colorado and Two Ocean Mountain in Wyoming. Also known from AA habitats in Europe. August.

Notes The dark cap contrasts with the bright gills and white stipe. *C. adelbertii*, from subalpine conifer forests, including those in Colorado, Wyoming, and Washington, might be the same. The alpine form *C. adelbertii* f. *alpina* described by M.M. Moser from Wyoming fits our fungus.

- Cap umber-brown, smooth
- Gills bright orange; stem brown, covered with white fibrils
- Odor earthy or fragrant
- With dwarf willows

References Moser 1993; Moser et al. 1995. CLC 1659, sequenced.

Cortinarius vibratilis Fr.

Cap 2–2.5 cm across, convex, slightly domed, golden yellow, yellow-ocher, smooth, viscid, with white rim. **Gills** adnate to subdecurrent, pale buff with white floccose edges. **Cortina** existing as white tissue at cap margin. **Stem** 3 × 0.4 cm, narrow at apex, clavate swollen base, and then tapered at very base, dingy whitish ocher, tacky. **Flesh** whitish with ocher tints. **Odor** not distinct. **Taste** cap slime bitter. **Spores** rusty, 7–8 × 5–6 μm, elliptical, weakly verrucose.

Ecology and Distribution In the alpine with shrub willow *Salix glauca* on Independence Pass, Colorado. Typically known as a forest species at lower elevations in Europe and North America. August in the alpine.

Notes Except for *Cortinarius alpinus* and *Cortinarius absarokensis*, *Myxacium* species are rare in the alpine. *Cortinarius vibratilis* has a bright yellow, slimy cap and slimy white stem for easy recognition. *Cortinarius pluvius* is a separate species.

- Cap bright yellow and viscid
- Gills pale; stipe whitish ocher, clavate, sticky
- Taste bitter
- With shrub willow

Reference Knudsen and Vesterholt 2008. CLC 1762, sequenced.

Cortinarius hinnuleus var. *faveanus*.

Cortinarius fuscoflexipes CLC 3405.

Entoloma in Alpine and Arctic Habitats

Entolomas can be prolific in alpine and Arctic habitats at certain times and in certain places, mostly in mats of dwarf willows. Entolomas are morphologically variable and highly diverse, making recognition to genus sometimes challenging. The gills of young fruiting bodies are often white or gray; when gills and spores turn pink at maturity, fruiting bodies are more easily recognized as Entolomas. With the microscope, angular spores confirm the genus. Other genera with pink spores are rare in these cold climates.

Over thirty species of *Entoloma* are reported from Arctic and alpine habitats, and seven are reported here from the Rocky Mountains. Entolomas vary from tiny and delicate to robust and stocky. Cap colors are usually in the blackish brown, grayish brown to yellow-brown range for alpine species, and caps are usually smooth but sometimes striate. Stems are typically whitish to gray and smooth in larger species; but a bluish gray hue is possible for some smaller types. Odors are important to note, some possibilities being nitrous (chlorine), farinaceous (mealy), rancid, or fishy.

Most of the larger types appear to be mycorrhizal with willows or at least associated with them. Smaller types are likely decomposers. Some species appear restricted to Arctic-alpine habitats.

Entoloma sericatum CLC 3576.

Key to Entoloma in Alpine and Arctic Habitats

Spore print pink, spores angular; gills white, pink, grayish.

1. Small fragile types; cap striate to slightly striate 2
1. Larger fleshy types; cap smooth or fibrous 3

2. Cap center sunken, yellow-brown; stem pale **Entoloma aff. aurantioalpinum**
2. Cap convex, medium brown; stem bluish gray **Entoloma cf. montanum**

3. Cap conic-convex, with pointed apex; stem thin, gray-brown, fibrous 4
3. Cap convex or irregular convex; stem fleshy, white or gray-brown, smooth 5

4. Cap uniformly dark brown, pointed, rather smooth **Entoloma vernum**
4. Cap gray, fibrous, fragile, splitting **Entoloma sericeoalpinum**

5. Stem gray-brown; cap blackish brown, gills grayish brown **Entoloma atrosericeum**

5. Stem white 6

6. Cap uniform blackish brown; stem stout, clavate **Entoloma alpicola**

6. Cap medium brown; stem long, equal **Entoloma sericatum**

Entoloma alpicola (J. Favre) Bon & Jamoni

Cap 2–5 cm wide, fleshy, wavy convex, slightly conic-convex, with low, round umbo or not, uniformly very dark brown, drying lighter brown, smooth, greasy; margin turned in at first. **Gills** deeply adnexed, becoming broad, white, then pale pink, pink; edges whiter. **Stem** 1.5–3.0 × 0.8–1.5 cm, stout, clavate, white, watery brown in old collections, smooth-looking but pruinose for whole length. **Flesh** white. **Odor** farinaceous (mealy). **Spores** pink, 8–10 × 7–9 μm, angular, 5-sided.

Ecology and Distribution Common in the Rocky Mountain alpine from Colorado to Montana, in mats of dwarf willow and *Dryas*, often in clusters. Reported from Arctic and alpine habitats in Europe, Svalbard, and Iceland with dwarf willows. August.

Notes A robust species with dark brown fleshy cap and white clavate stem. Sequences match the *Entoloma alpicola/majaloides* complex. *Entoloma majaloides* has an ocher cap and is reported from other habitats. Either the ITS region is not sufficient to separate the two species, or they are the same. In the alpine, *E. sericatum* usually has a longer stem and is less compact.

- Cap fleshy, uniformly dark brown
- Gills pinkish; spores pink
- Stem white, clavate or tapered
- Odor farinaceous
- In willows or *Dryas*

Reference Brandrud et al. 2018. CLC 3548, Genbank PQ192574.

Entoloma atrosericeum Kühner (Noordel.)

Cap 1.5–3.5 cm, fleshy, convex to almost flat, with or without a depressed center, deep umber brown to almost blackish, smooth, not hygrophanous, greasy, not striate, not splitting. **Gills** deeply adnexed, somewhat separated, grayish brown, brown, or with pinkish hue. **Stem** 2–4 × 0.3–1.0 cm, rather flat, equal or slightly larger or tapering at base, watery dark brown, gray-brown, with scattered lighter fibrils. **Flesh** watery grayish brown. **Odor** farinaceous (mealy). **Spores** pink, 8–10 × 7–8 µm, isodiametric, weakly angular. Cap hyphae with incrustations.

Ecology and Distribution On the Beartooth Plateau in Montana and Wyoming in the North Central Rocky Mountains with dwarf willows *S. reticulata* and *S. arctica,* and also shrub willows. Often clustered. Reported from the Pyrenees, the Alps, and Greenland. August.

Notes A very dark, almost black, robust species with a dark stem. Our ITS sequence is closest to the type of *E. atrosericeum.* This name previously has been misapplied to some collections of *E. sericeoalpinum.*

- Cap fleshy, deep brown to almost blackish, smooth
- Gills grayish brown with pink hue; spores pink
- Stem grayish brown
- Odor farinaceous
- With willows

Reference Vila et al. 2020 (2021). CLC 3880, sequenced.

Entoloma* aff. *aurantioalpinum Armada, Vila et al.

Cap 2.5–3.0 cm wide, shallow convex, with sunken center, medium brown, yellow-brown, umber in center, radially striate in age, with fine appressed scales. **Gills** deeply adnexed, well separated, a bit thick, light or dark pink; edges concolorous. **Stem** 1.8–2.2 × 0.2–0.4 cm, equal, light yellowish brown with fine black hairs to base (use hand lens); base white mycelioid. **Flesh** thin in cap. **Odor** Fragrant. **Spores** pink, 11–12(–14) × 8–9 μm, angular, long.

Ecology and Distribution Rare on the Beartooth Plateau and on Independence Pass in Colorado among *S. reticulata*. Also known from the alpine in France, Italy, and Austria with *Salix* and *Dryas*. August.

Notes *Entoloma aurantioalpinum* was recently described from alpine areas of the Alps along with another taxon that will be described later. Our taxon molecularly matches the undescribed species, giving it an intercontinental distribution.

- Cap yellow-brown, depressed, striate, minutely scaly
- Gills pink, well separated; spores pink, big
- Stem light yellow-brown, smooth; with fine black hairs
- With willows

References Buyck et al. 2022; Noordeloos 2022. CLC 3562, sequenced; also CLC 1764, CLC 1374.

Entoloma cf. *montanum* Noordel. et al.

Cap 0.8–2.5 cm across, convex indented in center, somewhat lobed, medium brown, covered with fine brown hairs, indistinctly striate. **Gills** attached, well separated, thickish, white, then pink; edges concolorous. **Stem** 1–3 × 0.1–0.3 cm, equal, flattened with a groove, deep steel blue-gray, gray, white at base, smooth, polished. **Flesh** whitish. **Odor** indistinct. **Spores** pink, 9–11 × 6–9 μm, angular.

Ecology and Distribution Rare in the alpine of Colorado, Wyoming, and Montana with dwarf willow *S. reticulata*, or mixed *Betula* and *Salix* species, sometimes on rocky cliffs. *E. montanum* is confirmed from the Caucasus, Alps, and Pyrenees in Europe with willows and *Dryas*. July and August.

Notes *Entoloma montanum* was recently described from alpine areas of Europe in snowbeds and heaths. Morphologically and ecologically, it fits our fungus, but molecular confirmation is needed. It keys out in section *Cyanula,* the blue-stemmed species.

- Cap small, convex, brown, slightly striate
- Gills white then pink; spores pink
- Stem deep steel blue-gray, polished, flattened
- With willows, possibly birch

Reference Noordeloos 2022. CLC 1981, CLC 1623, CLC 1396.

Entoloma sericatum (Britzelm.) Sacc. complex

Cap 1.5–5(!) cm wide convex, or very slightly conic-convex, to almost flat with low umbo, uniform dark brown, drying lighter, smooth, greasy, not striate. **Gills** sinuate, pale whitish gray at first, then pale pink, pink. **Stem** 3–6(!) × 0.5–1.5 cm, equal, or tapered at base, white with a grayish ocher sheen, silky, fibrous, floccose at apex. **Flesh** white, rubbery. **Odor** indistinct. **Spores** pink, 7–8 × 6–8 μm, angular.

Ecology and Distribution Rare in the North American alpine, reported from Montana/Wyoming with shrub willow *S. planifolia* and dwarf willow *S. reticulata*. Known in Europe from deciduous forests and the Arctic-alpine. August in the alpine.

Notes A large, tall *Entoloma* with a dark brown cap and long white stem. Molecular data places it in the *Entoloma sericatum* clade. *E. subarcticum* is likely synonymous; *E. svalbardense* is closely related. *Entoloma alpicola* is another large Arctic-alpine *Entoloma* that is shorter and stouter with a farinaceous odor that is not related.

- Cap brown, rather smooth-fibrous
- Gills gray to pale pink; spores pink
- Stem long, white, and fibrous
- Odor not distinct
- With willows

References Kokkonen 2015. Brandrud et al. 2018. CLC 3576, Genbank PQ187421.

Entoloma sericeoalpinum Vila, P.A. Moreau, Corriol & Reschke

Cap 1–3 cm wide, conic-convex, convex, with pointed umbo or not, grayish brown, lighter on drying, radially fibrous, smooth-silky, fragile, splitting, with tiny glistening particles. **Gills** adnate, a bit separated, pale to dark pink; edges lighter. **Stem** 1.5–4 × 0.2–0.4 cm, long and thin, cylindrical, pale brown, and sparsely pruinose in top part, white and fibrous in lower part. **Flesh** brittle, watery whitish. **Odor** possibly farinaceous or rancid. **Spores** pink, 9–12 × 8–10 μm, almost isodiametrical, with weak angles.

Ecology and Distribution In the Rocky Mountain alpine with willows or *Dryas*, often in clusters. Also from alpine areas of Spain, Italy, France, and Greenland. August.

Notes *Entoloma sericeoalpinum* is the alpine version of *E. sericeum*. Both species have a pointed fragile cap that tends to split, a brittle stem, and often a rancid odor. *E. sericeoalpinum* occurs in alpine situations with willow and *Dryas*, while *E. sericeum* is common at lower elevations.

- Cap conic-convex, convex, grayish brown, fragile, splitting
- Gills pink; spores pink
- Stem light brown to whitish, fibrous, brittle
- Odor often rancid
- Near willows or in open meadows

References Vila et al. 2020 (2021); Noordeloos 2022. CLC 3563.

Entoloma vernum S. Lund.

Cap 1.5–3.5 cm wide, strongly conic-convex with pointed umbo, uniformly medium grayish brown, drying lighter, smooth, dry, a bit shiny, with tiny sparkles, not striate. **Gills** deeply adnexed, buff, then pinkish gray-brown; edges concolorous. **Stem** 2–3 × 0.2–0.3 cm, long and thin, equal, dingy cream, then pale grayish brown, not distinctly pruinose, dry, fibrous with a few white fibrils. **Flesh** buff. **Odor** not distinct. **Spores** pink, 9–12 × 8–10 µm, nearly isodiametrical, weakly angular; caulocystidia present, cylindrical; cheilocystidia absent.

Ecology and Distribution This taxon is reported from Alaska and the Beartooth Plateau for AA habitats. *E. vernum* proper is rare in the alpine, and more common in spring conifer forests in Europe and western North America. July to August in the alpine.

Notes The uniformly brown pointed cap, pruinose stem, and cold habitat are the best identifiers for this mushroom. It is most common around melting snowbanks in spring in lower elevation forests. *Entoloma cuculatum* J. Favre is possibly the same species and occurs in the Swiss alpine with dwarf willow. Our spore shape fits that of variety *isodiametrica* described by Largent.

- Cap conic-convex, grayish brown, sparkly
- Gills pink; spores pink; stem grayish brown, thin
- Odor not distinct
- Usually near alpine willows

References Noordeloos 1992, 2022; Largent 1994. CLC 3589b, sequenced.

Hebeloma hiemale.

Hebeloma in Arctic and Alpine Habitats

A high diversity of *Hebeloma* species live in Arctic and alpine habitats and are mycorrhizal with shrub and dwarf willows, more rarely *Dryas*. Many species are difficult to differentiate from each other. Caps are convex or flat in the center, smooth, and greasy. Cap colors are limited to shades of cream, buff, brown, grayish brown, or brown, sometimes with a hoary coating. The cap rim can be much lighter, giving some species a two-toned look. Gills are white to light coffee color when young, which contrasts with the brown spore print. Stems are typically white and fibrous or rough. Most species have an odor of radish, raw potatoes, or cocoa. Training your nose to detect this odor can mean instant recognition of the genus after just a sniff or two, although some *Cortinarius* species also have the raphanoid odor. Some Hebelomas have a cobwebby veil that runs from the edge of the cap to the stem. Others lack a veil and instead have beaded drops of liquid along gill edges. Microscopically, Hebelomas have brown, elliptical to almond-shaped spores that are smooth to slightly bumpy. Cheilocystidia are consistent on gill edges. The number of gills is an important feature given here for each species. It can be quickly estimated by counting a quarter of the gills and multiplying by four.

Other small mushrooms in the alpine with brown spores include Inocybes, which usually have a more textured or fibrous cap, and an odor of burnt sugar, sperm, fish, or green corn. Small Telamonias (*Cortinarius*) have smooth caps that are usually pointed and stems with tell-tale bands of white tissue.

Key to Hebeloma in the Alpine and Arctic

WITHOUT VEIL (cortina)

1. Cortina (veil) absent (check young specimens) 2
1. Cortina (veil) present; stem base often dark 7

2. Cap small 1–2 cm; stem 0.2–0.4 cm wide; 20–40 gills 3
2 Cap larger 2–6 cm; stem 0.5–1.5 cm wide; 40–100 gills 5

3. Cap buff, brown, reddish brown; spores verrucose
   Hebeloma vaccinum
3. Cap a different color; spores smooth to slightly bumpy 4

4. Cap grayish brown, hoary **Hebeloma subconcolor**
4. Cap pinkish buff, orange-brown, with crenate white rim
   Hebeloma aurantioumbrinum

5. Cap buff, brown; stem pruinose; in *Dryas* or willows
   Hebeloma alpinum
5. Cap cream, buff, yellow-buff, pinkish buff; stem robust,
   floccose 6

6. Spores verrucose; mostly with dwarf willows **Hebeloma hiemale**
6. Spores a bit rough; with *Dryas* and shrub willows
   Hebeloma velutipes

WITH VEIL (CORTINA)

7. Medium types; stem 0.4–0.8 cm wide; spores ellipsoid,
   rather smooth 8
7. Smaller types; stems 0.1–0.5(–0.8) cm wide; spores
   almond-shaped, rough 10

8. Cap ocher, darker in center, 2-toned **Hebeloma mesophaeum** & others

8. Cap rather dark: brown, reddish brown, often hoary 9

9. Cap dark brown, hoary, with copious veil remnants; scattered **Hebeloma marginatulum**

9. Cap reddish brown; shiny, veil scant; often clustered **Hebeloma alpinicola**

10. Cap 1–2.5 cm, center blackish brown; stem 1–4 mm wide; often in moss **Hebeloma nigellum** & others

10. Cap 2–4 cm, brown, hoary; stem 3–8 mm wide **Hebeloma discomorbidum**

Non-veiled *Hebeloma* with droplets on the gills.

Hebeloma alpinicola A.H. Smith, V.S. Evenson & Mitchel

Cap robust, fleshy, 2–4 cm across, irregular convex, domed or not, reddish brown center with gray tones, lighter at margin, with hoary coating that dries shiny; margin turned down. **Cortina** present. **Gills** toothed or pulling away, somewhat broad, milk coffee, 36–44 gills reach the stem; edges floccose. **Stem** 3–4 × 0.5–1 cm, equal, whitish and pruinose at apex, dingy ocher and fibrillose in lower part; base darker, sometimes encased in sand or earth. **Flesh** dingy whitish, darker in stem. **Odor** raphanoid. **Spores** brown, 8–11 × 5–6 µm, ellipsoid, smooth to slightly rough, not dextrinoid. Cheilocystidia cylindrical but swollen near base.

Ecology and Distribution In the low alpine or at treeline, in the Rocky Mountains with willows, *Dryas,* or *Persicaria*. Widely distributed, occurring in western North America and Canada in boreal forests and in the Faroe Islands, Finland, Iceland, Norway, Scotland, Sweden, and alpine Austria, France, Italy, Romania, and Switzerland. Gregarious to cespitose. August.

Notes Unlike other robust Hebelomas found in the alpine, *Hebeloma alpinicola* is reddish brown and has a cortina. It is related to the *mesophaeum* group because of small ellipsoid, non-dextrinoid spores and the cystidia type. Compare with the darker *H. marginatulum*.

- Cap robust, reddish brown, shiny
- Gills pale, over 40; stem fibrous, dark at base
- Cortina present; spores smooth, not dextrinoid
- Gregarious to clustered
- With willows, and possibly *Dryas* and conifers

References Cripps et al. 2019; Eberhardt et al. 2023. CLC 1577, Genbank MK281065.

Hebeloma alpinum (J. Favre) Bruchet

Cap 2–3.5 cm across, convex to broadly domed, buff to pale brown, brown, paler at margin but not 2-toned, smooth, cracking when dry; margin turned under. **Cortina** absent. **Gills** attached, emarginate, pale milk coffee, 40–70; edges white fimbriate, beaded. **Stem** 1.5–3 × 0.4–1 cm, rather short, equal or clavate, white, firm. **Flesh** buff. **Odor** slightly raphanoid. **Spores** brown, 10–12 × 6–7 µm, almond-shaped with snout, slightly rough, slightly dextrinoid; cheilocystidia clavate-stiptate.

Ecology and Distribution *Hebeloma alpinum* appears confined to Arctic-alpine habitats and is reported from the Alps, Carpathians, Pyrenees, Greenland, Iceland, Scandinavia, Svalbard, Switzerland, Canada, and the Rocky Mountains. With willows, *Dryas,* and *Persicaria,* in August.

Notes A relatively robust alpine species without a veil. The more common *H. hiemale,* which has more verrucose spores, is often mistakenly called *H. alpinum. H. velutipes* has a robust floccose white stem.

- Cap robust, pale brown, brown, cracking in age
- Cortina absent
- Over 40 gills, with beaded drops
- Stem fleshy, short
- With *Salix, Dryas, Persicaria*

References Cripps et al. 2019; Beker et al. 2016. CLC 2855, sequenced (not above), Genbank MK281073.

Hebeloma aurantioumbrinum Beker, Vesterh. & U. Eberh.

Cap small, 1–2 cm across, convex, slightly conic-convex, smooth, greasy, cream, buff, pinkish buff, orange-brown, lighter at margin, not two-toned, hoary; margin turned in, crenate with a white rim. **Cortina** absent. **Gills** deeply indented, less than 40 reach stem, cream, buff, pinkish buff, milk coffee, brown; edges white fimbriate, drops visible. **Stem** 1.5–3 × 0.2–0.3 cm, equal, dingy whitish cream, darkening at base to watery brown, floccose/pruinose at top, smooth-fibrous below. **Flesh** dingy whitish. **Odor** faint or raphanoid. **Spores** yellowbrown, 10–13 × 6–7.5 µm, almond-shaped, slightly bumpy. Cheilocystidia clavate.

Ecology and Distribution In Colorado, Montana, and Wyoming with shrub willows *Salix glauca, S. planifolia,* and dwarf willow *S. arctica*. Occurring in Arctic-alpine habitats in the Northern hemisphere, including Svalbard, Iceland, Norway, and Sweden with willows. August.

Notes This small species has likely been mistaken for *H. pusillum,* which is slimmer, 2-toned, and typically at lower elevations in scrublands in warmer climates. Molecularly, the ITS region is almost identical to that of *H. helodes* Favre. Also, see *H. vaccinum*. There are few tiny Hebelomas that lack a veil.

- Cap small, pinkish buff; margin crenate; white rim
- Veil absent
- Fewer than 40 gills; beaded drops possible
- With willows

Reference Cripps et al. 2019. CLC 3093, Genbank MK281075.

Hebeloma discomorbidum (Peck) Peck

Cap 1.5–3 cm across, convex, not umbonate, smooth, dry or greasy, medium brown, bay brown, reddish brown, dark blackish brown, with hoary coating; white veil remnants at margin. Thick waxy pellicle in one collection. **Cortina** present. **Gills** emarginate, subdistant, 40–50 reach the stem, cream, milk coffee, pinkish cinnamon; margin floccose, white. **Stem** 1.5–6 × 0.3–0.8 cm, equal or enlarged at base, whitish, buff, brown, blackish brown at base, with patches of fibrils. **Flesh** watery buff, blackish brown in base. **Odor** raphanoid. **Spores** pale brown, 10–14 × 6–8 µm, almond-shaped, rough, dextrinoid. Cheilocystidia subcapitate with long neck, gradually swollen base.

Ecology and Distribution In Montana and Colorado, in the low alpine, primarily with shrub willow *S. planifolia*. Also occurs in alpine and Arctic habitats in Canada, Greenland, Scandinavia, Svalbard, and the Carpathians with willows. August.

Notes First described from New York bogs. Closely related and similar to *H. nigellum,* which has fewer gills. *I. oreophilum* and *I. palustre* are synonyms.

- Cap blackish brown with hoary coating
- Veil present and persistent
- Over 50 gills
- Stem with white fibrils, blackish brown base
- With willows

References Eberhardt et al. 2022; Cripps et al. 2019. CLC 3607, Genbank MK281072.

Hebeloma hiemale Bres.

Cap 1.5–3.5 cm in diameter, slightly conic-convex or domed-convex, smooth, greasy, pinkish buff, yellowish buff, to pale cream at the margin, with uniform coloration, somewhat hoary; margin turned down, then wavy. **Cortina** absent. **Gills** narrowly attached, emarginate, 48–60, white, milk coffee, pale brown, wood brown; edges white floccose, with drops of liquid. **Stem** 2–4.5 × 0.5–1.2 cm, equal, slightly clavate, whitish cream, totally pruinose (big floccules). **Flesh** whitish, firm. **Odor** raphanoid, faint. **Spores** yellowish brown, 10–12 × 6–7 μm, fat-bellied almond-shaped, lemon-shaped, distinctly verrucose, not or weakly dextrinoid; cheilocystidia clavate with thickened middle.

Ecology and Distribution Common in the Colorado, Montana, and Wyoming alpine. In Arctic and alpine zones in the Northern Hemisphere with dwarf willows, *Dryas* and birch. Also known in Europe from lowland dunes, shrublands, gardens, parks, and forests. August in the alpine.

Notes Many North American collections previously labeled *H. alpinum* are now confirmed as *H. hiemale,* which has a slimmer stipe and more warty spores. The brown on the cap in the photo shows the spore color.

- Cap fleshy, pinkish buff, yellowish buff
- Veil absent
- Over 40 gills reach the stem; liquid drops possible
- Stem robust, with floccules
- With willows, *Dryas*, birch

References Beker and Eberhardt 2010; Cripps et al. 2019. CLC 3094, CLC 3574, Genbank MK281028.

Hebeloma marginatulum (J. Favre) Bruchet

Cap 1.5–4 cm across, slightly conic-convex, sometimes flat or dished in center, smooth, shiny, strongly hoary, underneath dark brown, dark chestnut, dark caramel, mostly uniform, with fine white border near cap edge; margin turned in, covered with copious veil. Cuticle sometimes thick and rubbery. **Cortina** present, remnants distinct. **Gills** emarginate, 30–40, cream, pinkish buff, medium brown; edges fimbriate. **Stem** 2–4 cm × 0.2–0.8 cm, equal, pale buff, dark at base, apex pruinose, fibrous below. **Flesh** whitish, dark in base. **Odor** raphanoid, sourish, or faint. **Spores** yellowish gray, 10 × 6.4 μm, ellipsoid, smooth to slightly rough, not dextrinoid; cheilocystidia with long equal neck, somewhat swollen at base.

Ecology and Distribution In the Rocky Mountain alpine, with dwarf willows *Salix arctica* and *S. reticulata*. Restricted to AA habitats with dwarf willow; confirmed in Canada, Greenland, Iceland, Scandinavia, Svalbard, the Alps, and the Carpathians. Common in August.

Notes Often reported in AA habitats as *H. bruchetii,* which has been found to be *H. mesophaeum,* which has smaller spores.

- Cap uniformly brown, with hoary sheen
- Gills buff to brown; 30–40 in number
- Veil copious, leaving remnants
- Stem 2–4 mm wide, darker at base
- With dwarf willows

References Beker et al. 2016; Cripps et al. 2019. CLC 3545, Genbank MK281070.

Hebeloma mesophaeum (Pers.) Quél.

Cap 1–2 cm across, convex with low umbo, smooth, shiny, greasy, yellowish brown in center, outward lightening to pale ocher, at margin buff, 2-toned; margin turned in when young. **Cortina** present. **Gills** attached, adnate, 38–40, pale buff, pinkish buff, pinkish brown; edges fimbriate. **Stem** 3–4.5 × 0.3–0.6 cm, very gradually larger at base, white, pruinose at apex, fibrillose and darker to blackish at base. **Flesh** pale, dark in stipe base. **Odor** raphanoid. **Spores** yellow-brown, 8–11 × 5–6.5 µm, ellipsoid, almost smooth, not dextrinoid; cheilocystidia cylindrical with a swollen base.

Ecology and Distribution In the Colorado alpine with shrub willow *Salix glauca*. Common in the Arctic-alpine, also in boreal and subalpine habitats, with a wide range of hosts, including willows. August.

Notes Previously, *Hebeloma bruchetii* Bon was one of the most reported species from Arctic and alpine areas, but it is synonymized with *H. mesophaeum*. The closely related *H. excedens, H. alpinicola,* and *H. dunense* (=*H. velatum*) are similar and best distinguished molecularly. This group has small elliptical, smoothish, non-dextrinoid spores and cylindrical cystidia with a swollen base.

- Cap 2-toned: yellow-brown center, pale margin
- Veil present; gills pinkish buff
- Stem 3–6 mm wide, darker at the base
- With willow in alpine, but other hosts are possible

References Beker et al. 2016; Cripps et al. 2019. CLC 1245, Genbank MK281105.

Hebeloma nigellum Bruchet

Cap 1–2 cm across, convex with a low umbo, greasy, smooth, in center dark date brown, chocolate brown, or blackish brown, at margin cream, appearing 2-toned, with hoary sheen, glazed-looking; margin inrolled at first. **Cortina** present. **Gills** emarginate, L = 24–32, whitish, pale milk coffee; edges floccose. **Stem** 1.5–5 × 0.15–0.4 cm, long and slim, equal, undulating, dingy whitish and pruinose at top, blackish brown at base and silky-shiny. **Context** whitish, brown in base, rubbery in stipe. **Odor** raphanoid. **Spores** yellowish brown, 10–14.5 × 6–8 µm, almond-shaped, slightly rough, dextrinoid; cheilocystidia base swollen.

Ecology and Distribution Widespread across northern Europe, not only in Arctic-alpine habitats; reported from AA habitats in Canada, Greenland, Iceland, Svalbard, and the European Alps. In the Rocky Mountains, mostly with shrub willow *Salix planifolia,* often in moss; reported from Colorado and Montana in August.

Notes *Hebeloma kühneri* Bruchet, often reported from AA habitats, is the same. The closely related *H. hygrophilum* (= *H. paludicola*) and *H. spetsbergense* are distinguished molecularly, although the former is usually found in boreal habitats.

- Cap small, brownish black center; margin lighter
- Veil present
- Gills pale; 24–32 in number
- Stem long and slim, 1–4 mm wide, darker base
- With shrub willows

References Beker et al. 2016; Cripps et al. 2019. CLC 3614, Genbank MK281071.

Hebeloma subconcolor Bruchet

Cap 1.5–2 cm, convex, with or without a low broad umbo, becoming plane, smooth, moist, light to medium brown, with a grayish tint or sheen, lighter at margin. **Cortina** absent. **Gills** adnexed, well separated, 25–30 reach stem, dull light brown; edges lighter. No beaded drops reported. **Stem** 1.5–3 × 0.3–0.4 cm, equal, brown, covered with longitudinal white fibers; apex lighter and pruinose. **Flesh** buff. **Odor** astringent. **Spores** yellowish brown, 10.5–12.5 × 6.5–7.5 μm, almond-shaped, slightly rough, dextrinoid; cheilocystidia clavate.

Ecology and Distribution Reported under willows at alpine elevations of 4000 m in Colorado, cespitose to gregarious. Known from Arctic and alpine locations in the European Alps, Greenland, Iceland, and Scandinavia. Fruiting in August.

Notes This small brownish species has a grayish cast to the cap and a dark stem; the gills are well separated. It should be compared to the other non-veiled, small species such as *H. aurantioumbrinum* and *H. vaccinum*.

- Cap small, brown with a grayish hoary cast
- Veil absent; gills well separated, 25–30
- Stem brown with white fibers
- With willows

References Beker et al. 2016; Cripps et al. 2019. CLC 1651, Genbank MK281117.

Hebeloma vaccinum Romagn.

Cap 1 cm across, convex, buff to brown with hoary coating, rather unicolor, smooth, shiny, tacky; margin turned down, a bit crenulate, faintly striate; edges white. **Cortina** absent. **Gills** adnexed, about 38 reach the stem, buff to milk coffee. **Stem** 1 × 0.3 cm, equal, cream, finely floccose at apex, fibrillose for length, delicate. **Flesh** cream. **Odor** faintly raphanoid. **Spores** yellowish brown, 10–14 × 6–8 µm, almond-shaped, lemon-shaped, verrucose, dextrinoid; cheilocystidia clavate, swollen at apex.

Ecology and Distribution In North America, recorded from the Colorado Rocky Mountains with shrub willow *Salix arctica*. In Arctic and alpine habitats, confirmed in the European Alps, the Carpathians in Slovakia, and Greenland, always with *Salix*. Also, in low-elevation dunes and woodlands with *Salix* and aspen (*Populus*) in Europe. August.

Notes This species is usually larger (up to 4 cm) than the specimens described here. In the Rocky Mountain alpine, it can be recognized by its association with dwarf willows, small size, and lack of a veil. Compare with *H. aurantioumbrinum,* which has slightly bumpy spores, and *H. subconcolor,* which has a grayer cap. Rare.

- Cap small, buff to brown, unicolor
- Veil absent
- Gills buff, 38 in number; stem pale
- With dwarf willow

References Beker et al. 2016; Cripps et al. 2019. CLC 1881, Genbank MK281113.

Hebeloma velutipes Bruchet

Cap 2–6 cm across, convex, a bit domed, tacky to kidskin, smooth, nearly unicolor, very pale buff, pale salmon buff, with hoary coating. **Cortina** absent. **Gills** narrowly attached, sinuate, 50–75 reach stem, white, then milk coffee; edges white-floccose, with beaded drops on some. **Stem** 3–6 × 0.7–1.5 cm, robust, equal or slightly swollen at base, mostly floccose, or fibrous in lower part. **Flesh** whitish, thick in cap, firm in stem, stuffed/hollow. **Odor** raphanoid. **Spores** brown, 10–12 × 6–7 µm, almond-shaped, slightly rough, dextrinoid; cheilocystidia clavate.

Ecology and Distribution In the Rocky Mountain alpine, mostly with *Dryas*. Recorded in several Arctic-alpine habitats, particularly Svalbard with *Dryas octopetala* and *Salix polaris*. Common in Europe and North America at lower elevations with deciduous trees. August.

Notes *Hebeloma velutipes,* a larger *Hebeloma*, is occasionally found in AA situations; it is recognized by a pale cap, robust white stipe covered with floccules, and an association with *Dryas*. It can be confused with *Hebeloma alpinum* or *Hebeloma hiemale*, which typically have shorter, less floccose stems.

- Cap robust, fleshy, pale buff, unicolor
- Veil absent
- Gills pale with floccose edges, beaded drops, over 40
- Stem robust, white, with floccules
- Usually with *Dryas*, possibly shrub *Salix*

References Beker et al. 2016; Cripps et al. 2019. CLC 1651, Genbank MK281117.

Inocybe in Alpine and Arctic Habitats

A large diversity of *Inocybe* species inhabit alpine and Arctic habitats, so they can be overwhelming to sort out. Many can be recognized by typical *Inocybe* features such as small, brown, fibrous, pointed caps (fiber heads); brown spores; and a spermatic odor. Other species have white to light yellow-brown caps and odors that are fragrant, spicy, or absent. Other small-capped, brown-spored groups, such as *Hebeloma* and *Telemonia* (*Cortinarius*), typically have smoother caps, while those of *Inocybes* are often (but not always) fibrous, rough, or scaly. A challenge for field identification is that microscopic features define groups; these features are combined with macro-morphology here to foster tentative field identification, which should be confirmed by microscopic examination when possible.

For example, *Inocybe* species with smooth spores typically have a cortina, generally lack a marginate (rimmed) stem base, and only the top part of the stem is pruinose. Those with nodulose spores often lack a cortina (observe young specimens), often have a marginate stem base, and the stem is totally pruinose. A subgroup of these has spores that are barely nodulose and almost rectangular; these lack a pruinose stem and marginate base. To determine pruinosity in the field, avoid touching the stem while picking and hold the specimen up to the light; cystidia sparkle in sunlight (use a hand lens). There are exceptions to each generality. For example, *I. leiocephala* has smooth spores and a totally pruinose stem.

All species in genus *Inocybe* have pleurocystidia on the sides of the gills; they are mostly thick-walled and the shape can be diagnostic. Species lacking pleurocystidia are now in *Mallocybe, Inosperma*, and *Pseudosperma* (pg 168). *Inocybes* are mycorrhizal with willows, *Dryas*, and birch in alpine and Arctic habitats, although occasionally they pop up in moss.

Three Groups of Alpine and Arctic Inocybes

All species have pleurocystidia as thick-walled cells along sides of gills. If pleurocystidia are absent, see *Inosperma, Pseudosperma*, and *Mallocybe*, pg 168. First select the appropriate group and then use the key for that group to determine species.

1. Spores smooth; cortina usually present; stem pruinose only in top part or not; base not marginate (exception: *I. leiocephala* has a totally pruinose stem), pg 144
2. Spores nodulose; cortina lacking; stem totally pruinose; base marginate, pg 155 (base not marginate in one)
3. Spores angular-lumpy, rectangular; cortina ephemeral; stem not pruinose; base not marginate, pg 161

Group 1: smooth spores

Group 2: nodulose spores

Group 3: rectangular spores

Three Main Groups of *Inocybe*. Top Row. Group 1. Cortina present; stem equal, pruinose at top. Group 2. No cortina; stem marginate, totally pruinose. Bottom Row. Group 3: Cortina ephemeral; stem equal, not pruinose. Spore types for 3 groups.

INOCYBE GROUP 1: Smooth spores, cortina present; stem not marginate, apex pruinose or not

1. Cap white or pale yellow-brown; can be tinted lilac, pink, reddish or not 2
1. Cap more highly colored 4

2. Odor spicy; cap white to pale yellow-brown, reddening
   Inocybe fraudans
2. Odor spermatic 3

3. Cap white with pinkish stains in age **Inocybe pudica** (see *I. pallidicremea*)
3. Cap white, with lilac tints when young **Inocybe pallidicremea**
(if base bulbous, see *Pseudosperma bulbosissimum*)

4. Cap fibrous, fibrous-smooth 5
4. Cap rough-fibrous, scaly 7

5. Cap ocher brown, dark brown; stem totally pruinose **Inocybe leiocephala**
5. Not as above; stem pruinose at apex 6

6. Cap pale brown; stem covered with white fibrils **Inocybe maculipes**
6. Cap yellow-brown, brown; stem smooth or fibrillose **Inocybe nemorosa**

7. Cap dark brown, red-brown, rough; stem red-brown, rough; odor spermatic; spores almond-shaped **Inocybe rupestroides**
7. Cap dark brown, dark ocher, rough-fibrous, sub-scaly, scaly; stem pale, rough; odor not distinct or fungoid; spores long and narrow or ellipsoid 8
(If odor fishy, see *Inosperma calamistratum*)

8. Cap small, 0.5–2 cm, fibrous-rough; spores ellipsoid **Inocybe norvegica**
8. Cap larger, 1.5–3.5 cm, scaly; spores long and narrow 9

9. Cap margin regular **Inocybe lacera var. lacera**
9. Cap margin or cuticle strongly turned under **Inocybe lacera var. rhacodes**

(Note: The *Inocybe lacera* group is phylogenetically related to the nodulose-spored species, but is placed here because spores appear smooth)

Inocybe fraudans (Britzelm.) Sacc.

Cap 1.5–3.5 cm across, conic-convex, convex with umbo, white at first with covering, then pale yellow-brown, reddening in age, radially rough-fibrillose; margin with white tissue. **Gills** adnexed, cream, pink in some. **Cortina** as tissue on cap margin. **Stem** 2–4 × 0.4–0.6 cm, equal, cream, some with pink tint, longitudinally fibrous, pruinose for whole length. **Flesh** white, reddening; tough in base. **Odor** fragrant, spicy, cedar-cinnamon. **Spores** brown, 9–11 × 5–7 µm, smooth, fat almond-shaped with conic apex (lemon-like); pleurocystidia subfusiform.

Ecology and Distribution Occurring in the Rocky Mountain alpine with shrub *Salix* and *Dryas*. More common in conifer or birch forests. Known from other Arctic-alpine habitats, at least in the Alps, Greenland, and Norway. August.

Notes The pale textured cap, unique spicy odor and reddening flesh distinguish this species, which may be the same as *I. pyriodora*.

- Cap pale whitish, yellow-brown, reddening, rough
- Gills cream to pink
- Stem cream with pinkish flesh
- Odor fragrant, spicy, cedar-cinnamon; spores smooth
- With willows or *Dryas*

References Bon 1988; Knudsen and Vesterholt 2008. CLC 3115, CLC 1697 Genbank PQ187664.

Inocybe lacera var. ***lacera*** (Fr.) P. Kumm. complex

Cap 1.5–3.5 cm across, convex, sometimes with a low umbo, dark brown, center woolly, outwards fibrillose; margin turned down. **Gills** adnexed, buff with pinkish tint, then deep brown. **Cortina** in young specimens. **Stem** 1–2 × 0.1–0.3 cm, equal, buff then brown, darker at base, rough, not pruinose. **Flesh** pinkish in stem. **Odor** not distinct or fungoid. **Spores** brown, 11–15 × 4.5–6.5 µm, smooth, long cylindrical, some slightly angular; cystidia subfusoid, narrow, with pedicel.

Ecology and Distribution In many alpine habitats in the Rocky Mountains and Alaska with dwarf willow, shrub willow, or *Dryas*, often in open sandy or gravelly soil. Found in many Arctic-alpine areas outside North America, and at lower elevations in forests, often on sandy soil. August.

Notes This is a common species with a wide distribution. However, it is likely a complex of species that needs to be sorted out. Compare to *I. norvegica,* which is smaller and looks similar but has more elliptical spores. *I. heterospora* is also within this complex.

- Cap brown, rough-scaly
- Gills pale pinkish buff to brown
- Stem brown, not pruinose
- Odor fungoid or absent
- Spores long and narrow
- With willow or *Dryas*

References Armada et al. 2024; Kuyper 1986. CLC 1344.

Inocybe lacera var. *rhacodes* (J. Favre) Kuyper

Cap 1.5–3.5 cm across, somewhat bell-shaped with margin strongly turned under, yellow-brown, brown, with coarse recurved scales; cuticle overhangs cap edge and turns under. **Gills** adnexed, broad, cream then yellow-brown. **Cortina** buff. **Stem** 2–3.5 × 0.3–0.8 cm, stout in some, equal, dingy white at apex, blackish brown below, rough-fibrous, not pruinose. **Flesh** pink in stem. **Odor** weakly spermatic. **Spores** brown, 10.5–14 × 4.5–5.5 µm, long cylindrical, slightly angular; cystidia subfusiform with yellow-brown contents.

Ecology and Distribution Found on Independence Pass in Colorado, with willows. Reported in alpine areas of Switzerland and Norway with dwarf willows. August.

Notes This looks like a phenotypic variant of *I. lacera* that might be due to harsh conditions, but the taxon is distinctive because of a spermatic odor, brown cheilocystidia, and overhanging cuticle.

- Cap yellow-brown, scaly, with overhanging cuticle in some
- Gills cream, partly covered by cuticle in some
- Stem stout, brown, rough-fibrous, pink inside
- Odor spermatic; spores long and narrow
- With willows

Reference Kuyper 1986. CLC 1343, CLC 1331.

Inocybe leiocephala D.E. Stuntz

Cap 0.5–5 cm across, conic convex, campanulate, with umbo, pale reddish yellow, ocher brown, dark brown, smooth, somewhat shiny. **Gills** attached, well spaced, pale gray, gray-brown. **Cortina** not noted, margin clean. **Stem** 2–6(–8) × 0.3–0.7 cm, equal with subbulbous base or not, pale yellow-brown, sometimes reddish, often with whitish base, striate, pruinose for the whole length. **Flesh** whitish, can be reddish in stipe. **Odor** acidulous to spermatic. **Spores** brown. 8–10 × 5.5–6 µm, smooth, slightly almond-shaped to ellipsoid; cystidia fusiform.

Ecology and Distribution Common in the Rocky Mountain alpine with dwarf willows or *Dryas*. Known from other Arctic-alpine habitats, including Sweden, Finland, Svalbard, Greenland, and also from conifer forests in Washington state. August.

Notes Possibly the same as *I. catalaunica* Singer. The distribution from low-elevation conifer forests to Arctic and alpine habitats is unusual. One of few species with both smooth spores and a totally pruinose stem found in AA habitats.

- Cap ocher, dark brown, smooth, umbonate
- Gills pale gray-brown, well separated
- Stem pale, often reddish, striate, totally pruinose
- Spores smooth
- With dwarf willows or *Dryas*

Reference Larsson, Vauras, Cripps 2014. CLC 3777 (above); CLC 2325, Genbank KJ399911.

Inocybe cf. *maculipes* J. Favre

Cap 1.5–3 cm across, conic-convex, with small umbo, pale brown, brown, radially rough-fibrous, hoary, covered with fibrils when young. **Gills** toothed, pale brown, yellow-brown; edges floccose. **Cortina** white on cap margin. **Stem** 1.5–3.5 × 0.2–0.8 cm, equal or slightly larger toward base, watery reddish brown, orange-brown, covered all over with white fibrils, pruinose only at very apex or not at all. **Context** reddish brown in stem. **Odor** spermatic. **Spores** brown, 10–13(–17) × 5–7.5 µm, smooth, slightly almond-shaped to ellipsoid; cystidia subfusiform with long pedicel.

Ecology and Distribution On Independence Pass in Colorado with willows and *Dryas*. Also with *Dryas* in alpine Switzerland. August.

Notes The red-brown stem covered with white fibrils is distinctive, along with the other features. This high-elevation species is rarely recognized.

- Cap pale brown, brown, rough fibrous
- Gills pale brown or yellow-brown
- Stem strongly covered with white fibrils
- Odor spermatic; flesh reddish brown; spores smooth
- With willow or *Dryas*

References Favre 1955; Breitenbach and Kränzlin 2000. CLC 1340.

Inocybe nemorosa (R. Heim) Grund & D.E. Stuntz

Cap 1.5–2.5 cm across, conic-convex with sharp umbo, campanulate, pale yellow-brown, medium to umber brown, radially fibrous, fibrils can diverge, slightly greasy. **Gills** sinuate, well separated, pale brown, yellow-brown. **Cortina** uncertain. **Stem** 2–4 × 0.2–0.3 cm, long and thin, equal with slight bulb at base, smooth to fibrillose, watery orange, pruinose at apex. **Flesh** watery orange. **Odor** spermatic. **Spores** 9–10 × 6 μm, smooth, slightly fat almond-shaped; cystidia subfusiform with pedicel.

Ecology and Distribution Occuring in the San Juan Mountains of Colorado and the French and Swiss Alps with dwarf willows; first known from the conifer forests of Nova Scotia. August.

Notes *I. friesii* f. *nemorosa* is the same. This form has been reported from the Swiss and French Alps but is better known from the eastern forests of North America. Related to *I. nitidiuscula*.

- Cap campanulate, brown, fibrous-smooth
- Gills well separated, pale orange
- Stem and flesh watery orange
- Odor spermatic; spores smooth
- With willows or conifers

References Armada et al. 2024; Grund and Stuntz 1968; Favre 1955. Eastern subalpine version above JK029.3. Inset CLC 1694, Genbank PQ187664 (DNA matches type).

Inocybe norvegica Vauras & E. Larss.

Cap tiny, 0.5–2.0 cm across, hemispheric, convex, pale to dark brown, some with slight umbo, radially rough-fibrous. **Gills** adnate, pale brown. **Cortina** as a few buff fibrils. **Stem** thin, 1–2.5 × 0.2–0.4 cm, equal with ball of soil at base, light brown, darker towards base, rough-fibrous, not pruinose. **Flesh** buff. **Odor** fungoid or absent. **Spores** 10–12(–17) × 5–6.5 µm, most ellipsoid or slightly angular, highly variable; cystidia fusoid, mixed with pyriform, brown cells on gill edges.

Ecology and Distribution With shrub willows in gravel on Crow Pass near Girdwood, Alaska. Previously only known from Sweden and Austria with *Salix herbacea*. August.

Notes This new species is like a miniature *I. lacera*, but the microscopic and molecular data separate it. *I. lacera* var. *helobia* has different cystidia and spores. Likely overlooked or misidentified as *I. lacera*. Our spores are more variable than those of the type.

- Cap small, convex, brown, rough
- Gills buff
- Stem brown, darker at base, rough
- Spores ellipsoid, slightly angular
- With willows

Reference Vauras and Larsson 2021. CLC 3787, Genbank PQ191192.

Inocybe pallidicremea Grund & D.E. Stuntz

Cap 2–3 cm across, conic-convex, campanulate, with small umbo, lilac at first, fading to dark cream, light brown, rather smooth. **Gills** emarginate, well separated, buff with gray. **Cortina** as a few strands. **Stem** 2.5–3.5 × 0.3–0.4 cm, gradually enlarging toward base, cream with lilac tones, fibrous-smooth. **Flesh** cream, slightly lilac in cap and yellow in stem base. **Odor** spermatic. **Spores** 8–9 × 5–6 μm, smooth, mostly ellipsoid or slightly amygdaliform; cystidia subfusiform.

Ecology and Distribution In Alaska, with *Dryas,* willows, and birch in the low alpine. Primarily known from conifer forests in the western United States and Canada, and northeastern North America. August.

Notes This species has been misidentified as *Inocybe lilacina,* which is an eastern North American species. The lilac color can fade, suggesting *Inocybe geophylla. Inocybe pudica,* which turns pink, is also reported from the low alpine with willows from Alaska, but it is rare in this habitat.

- Cap cream, with or without lilac tints, smooth
- Gills pale brown
- Stem cream, with or without lilac tints
- Odor spermatic; spores smooth
- With willows, *Dryas*, and birch

References Grund and Stuntz 1977; Matheny and Swenie 2018. CLC 3791, Genbank PQ191193.

Inocybe rupestroides E. Larss. & Vauras

Cap 1–2.5 cm across, conic-convex, convex with small umbo, red-brown, yellow-brown, with dark center, rough-fibrous, excoriating. **Gills** adnexed, orange-brown with yellow tones. **Cortina** as a few pale fibrils. **Stem** 1.5–3.5 × 0.1–0.2 cm, long and thin, orange-brown, red-brown, darker in lower half, almost smooth or with fibrils. **Flesh** reddish in stem. **Odor** faintly spermatic. **Spores** brown, 9.5–12 × 5–6.5 µm, smooth, slightly almond-shaped; pleurocystidia spindle to flask-shaped with long necks, yellow contents, and thick walls.

Ecology and Distribution Typically cespitose under shrub willows. Known from the alpine in Colorado (Independence Pass) for North America. Also confirmed in Fennoscandia. August.

Notes The red-brown cap, stem, and flesh help define this newly described species, as does the roughened cap surface. Ours matches the type. *Inocybe auricomella* and *Inocybe rupestris* are similar, but *Inocybe rupestroides* becomes dark ochraceous brown and has larger spores.

- Cap red-brown, rough, drying lighter
- Gills orange
- Stem red-brown; flesh red-brown
- Odor spermatic; spores smooth
- With willows

References Larsson and Vauras 2023. Ferrari 2006. Bon 1988. CLC 1243, Genbank PQ191256.

Inocybe phaeocystidiosa CLC 3293, Genbank MK 153635 (subalpine).

INOCYBE GROUP 2: With nodulose spores, cortina absent, stems totally pruinose, with marginate base or not

1. Cap small (1–2 cm), convex, brown, paler toward margin; stem striate, without marginate base; flesh pinkish in stem **Inocybe baeocarpa**
1. Not as above; stem base marginate 2

2. Cap brown, campanulate with frosty velipellis **Inocybe arctica**
2. Cap yellow-brown, conic-convex, campanulate, without vellipellis 3

3. Cap appressed scaly; gills cream with yellow tint; base strongly marginate; odor faintly fruity; cystidia yellow brown **Inocybe phaeocystidiosa**
3. Not as above; cystidia pale 4

4. Cap smooth; gills gray; odor spermatic; spores 6–7 μm wide **Inocybe occulta**
4. Cap fibrous; gills gray-brown; odor absent; spores 7–8.5 μm wide **Inocybe albomarginata**

Inocybe alpinomarginata C.L. Cripps, E. Larss., & Vauras

Cap 1.5–3 cm, conic-convex, campanulate to flat, yellow-brown, smooth-fibrous, greasy. **Gills** sinuate, buff, pale gray, grayish brown. **Cortina** absent. **Stem** 1.5–4 × 0.2–0.4 cm, marginate to submarginate, white to watery orange, totally pruinose. **Flesh** white, sometimes yellow in base. **Odor** absent. **Spores** brown, 10.5–11.5 × 7.5–8.0 µm, nodulose; cystidia subfusiform, not yellow-brown.

Ecology and Distribution So far only known from the alpine of Colorado, Sweden, Italy, and Austria with dwarf and shrub willows. August.

Notes Similar to *I. phaeocystidiosa,* but in *I. alpinomarginata,* the cap is smoother, not so robust, and pleurocystidia are pale. *I. alpestris* is similar but spores have more nodules.

- Cap campanulate, yellow-brown, smooth-fibrous
- Gills grayish buff
- Stem marginate, white to pale orange
- Odor absent; spores nodulose
- With willows

Reference Cripps, Larsson, Vauras 2020. EL207–13B, Genbank MK153648 (type).

MYCORRHIZAL MUSHROOMS WITH WILLOW, BIRCH, *DRYAS*

Inocybe arctica E. Larss., Vauras & C.L. Cripps

Cap 1–3.5 cm, conical convex, convex with umbo, dark brown, rough at center, otherwise fibrous, often with frosty velipellis. **Gills** adnexed, somewhat broad, pale yellow then grayish brown with yellow tinge. **Cortina** scarce. **Stem** 1–3.5 × 0.3–0.7 cm, at base somewhat bulbous but not clearly marginate, yellowish with orange tinge, white at base, silky-shiny, pruinose in top half, but not evident in lower part. **Flesh** whitish buff, some reddish tint at top of stem. **Odor** absent or slightly spermatic. **Spores** brown, 11.5–12.0 × 7.0–8.0 μm, nodulose; cystidia slender.

Ecology and Distribution With dwarf willow *Salix arctica* in Colorado at 3800m; also from Sweden, Norway (type), and Svalbard (sea level) with dwarf willows *S. herbacea* or *S. reticulata*. July to August.

Notes The dark brown smoothish cap with a frosty covering and bulbous stem base help identify this species. It is close to *Inocybe favrei* Bon (= *I. taxocystis* (J. Favre) Singer).

- Cap dark brown, umbonate, with frosty covering
- Gills grayish brown
- Stem pale yellow-orange, silky-shiny, base bulbous
- Odor slightly spermatic; spores nodulose
- With willows

References Cripps et al. 2020; Larsson et al. 2018. CLC 1804, Genbank MK153628; El34–09x, Genbank KY033843 (type).

Inocybe baeocarpa E. Larss., C.L. Cripps, Bandini

Cap 0.8–2.5 cm, tiny, slightly conic-convex, or with low, broad umbo, dark brown in center, outward pale brown, velvety to rough; margin turned down with whitish rim. **Cortina** absent. **Gills** narrowly attached, white to cream or pale yellow; edges concolorous. **Stem** 8–15 mm × 1 mm, equal, a bit undulating, not marginate, pale cream with pinkish hue, pruinose to base. **Flesh** pinkish in top of stem. **Odor** subspermatic. **Spores** 8.5–10 × 6–7 μm, nodulose. Pleurocystidia bright yellow in KOH, with rounded base.

Ecology and Distribution Cespitose, in alpine with dwarf willow *Salix arctica,* shrub willow *S. glauca*, and krummholz spruce, also below treeline with mixed aspen-spruce, near the Beartooth Plateau, Montana. Also from alpine Sweden. August.

Notes Originally thought to be *I. subexilis* but separate according to molecular data. A tiny *Inocybe* with totally pruinose stem, brown cap, and pinkish stem flesh. The yellow cystidia are distinct.

- Tiny cap, brown in center, pale brown on margin
- Stem striate, whitish with pink
- Totally pruinose, but not marginate
- Spores nodulose; cystidia yellow in KOH
- With willows, maybe spruce

Reference Larsson et al. (in prep.). CLC 2856, CLC 3398, Genbank (in prep).

Inocybe occulta Esteve-Rav. et al.

Cap 1.5–2.5 cm across, conic-convex, campanulate with indistinct umbo, ocher, yellow-brown, a bit greasy, appearing smooth but can be scurfy in center and outward fibrillose. **Gills** sinuate, gray, grayish brown. **Cortina** absent. **Stem** 2–3 × 0.2–0.4 cm, equal down to marginate base, white, pruinose to the base, more obvious in top part. **Flesh** whitish. **Odor** spermatic when cut. **Spores** brown, 9.5–10 × 6–7 µm, nodulose, with one or two prominent nodules at the end; cystidia subfusiform.

Ecology and Distribution In alpine habitats with dwarf and shrub willows (Colorado), and in alpine and boreal habitats in Fennoscandia, but also possibly in North American and European forested habitats. August in the alpine.

Notes Most collections in alpine and Arctic habitats previously determined to be *I. mixtilis* (which occurs primarily in forests) are likely *I. occulta*. The gray gills, spermatic odor, and narrow spores separate it from similar AA species in this group.

- Cap yellow-brown, conic-convex, bell-shaped
- Gills gray
- Stem white, marginate, totally pruinose
- Odor spermatic; spores nodulose
- With willows

Reference Cripps et al. 2020. CLC 3695 (above, subalpine) matches CLC 1756 (alpine) 100%, Genbank MK153692.

Inocybe phaeocystidiosa Esteve-Rav., Moreno & Bon

Cap 1.5–3.5(–6.0 cm in subalpine collection), conic-convex, with rounded smooth umbo, golden brown, appressed scaly to fibrous, greasy. **Gills** sinuate, whitish cream, grayish brown, with yellow tints. **Cortina** absent. **Stem** 2–5 × 0.3–0.5 cm (up to 10 × 1 cm in subalpine collections), marginate and rounded at base, appearing smooth but totally pruinose (hand lens). **Flesh** white. **Odor** none or faintly fruity. **Spores** 10.5–11.5 × 7.5–8.5 µm, nodulose; cystidia somewhat utriform.

Ecology and Distribution In the alpine zone with dwarf willows *S. reticulata, S. arctica,* and shrub willow *S. glauca* in the Rocky Mountains (also in subalpine forests) and in alpine and forested areas of Europe. August.

Notes Alpine collections once called *I. praetervisa* are likely this species. The stem of *I. phaeocystidiosa* is entirely pruinose (not true of *I. praetervisa*), and the cystidia of the former sometimes have yellow-brown contents.

- Cap conic-convex, yellow-brown, appressed fibrillose
- Gills cream to gray
- Stem white, marginate, totally pruinose
- Odor none or faintly fruity
- Spores nodulose
- With willows

Reference Cripps et al. 2020. CLC 3107, Genbank MK153637.

Inocybe giacomi CLC 1700. Spores rectangular; stem base not marginate or pruinose.

INOCYBE GROUP 3: With rectangular-lumpy spores, cortina ephemeral, stem not pruinose, base not marginate

1. Cap tiny 10–20 mm, yellow-brown, scales with white tips **Inocybe leonina**
1. Cap brown, dark brown, purple-brown, gray-brown 2

2. Cap small (5–25 mm), with velipellis 3
2. Cap usually larger, without obvious velipellis 4

3. Cap dark brown, smooth to rough; gills rusty; odor absent **Inocybe purpureobadia**
3. Cap brown, gray-brown, fibrous to woolly; gills pale; odor spermatic **Inocybe murina**

4. Gills pale orange; cap convex, brown, gray-brown; stem without pink; odor absent **Inocybe paragiacomi**
4. Gills gray-brown; stem with pink tints; odor spermatic 5

5. Cap often tall conic-convex, campanulate, brown, gray-brown, streaky **Inocybe giacomi**
5. Cap convex or sunken in center, brown, gray-brown **Inocybe subgiacomi**

Inocybe giacomi J. Favre

Cap 1–3.5 cm across, shape variable, mostly tall conic-convex, with rounded umbo, campanulate, medium to dark brown, often with a gray tinge, even darkening to black, streaky, fibrous, smooth to rough, somewhat shiny. **Gills** sinuate, grayish brown, milk coffee. **Cortina** not observed. **Stem** 2–4.5 × 0.2–0.5 cm, equal, not marginate, pale orange-brown with pink tinge at top, fibrous, not pruinose. **Flesh** white, pink in stem. **Odor** subspermatic when cut. **Spores** brown, 9–11 × 5.5–7 µm, barely nodulose-rectangular; cystidia lageniform with pedicel.

Ecology and Distribution With shrub willow *Salix planifolia* and dwarf willows *S. arctica* and *S. reticulata* in North America (Colorado), and frequent in forested and Arctic-alpine areas of Europe. July to August.

Notes When the cap is tall, umbonate, streaky, and dark brown, it can be distinctive enough to help recognize this species. Compare to *I. paragiacomi* and *I. subgiacomi,* which lack the tall cap.

- Cap tall campanulate, streaky, dark brown
- Gills grayish brown
- Stem orange-brown with a pink tinge on top
- Odor faintly spermatic; spores barely nodulose
- With willows

Reference Cripps et al. 2020. CLC 1359, Genbank MK153653.

Inocybe leonina Esteve-Rav. & A. Caball.

Cap small, 1–2 cm across, conic-convex, hemispherical, golden, medium brown, smooth when young, then with uplifted scales with white tips. **Gills** sinuate, white, then grayish brown. **Cortina** not observed. **Stem** 1.5–2.0 × 0.3–0.5 cm, equal, squared off at base to indistinctly submarginate, shining white, pruinose to base but not distinct. **Flesh** white. **Odor** unusual, faintly fruity. **Spores** brown, 11.0–13.0 × 6.5–8.0 µm, barely nodulose; cystidia clavate, scarce.

Ecology and Distribution In the alpine zone of North America (Colorado, 3700m) where dwarf and shrub willows occur, also in the European subalpine (Spain, type). August in the alpine.

Notes Remarkably, this is only the second known collection of this species, and the type is from a mountainous locality in a pine forest of *Pinus sylvestris* at 1300 m in Spain.

- Cap golden yellow, developing scales with white tips
- Gills grayish brown
- Stem shining white
- Odor unusual, maybe faintly fruity
- Spores barely nodulose
- With willows

References Cripps et al. 2020; Larsson et al. 2018. CLC 1349, Genbank MG574396.

Inocybe murina E. Larss., C.L. Cripps & Vauras

Cap 1–2.5 cm across, conic-convex, convex with sharp umbo, medium to dark brown, woolly or smoother with a silvery white coating. **Gills** emarginate, white then pale grayish brown. **Cortina** soon gone. **Stem** 1.5–2.5 × 0.2–0.5 cm, equal, pale brown, pink tint in top half (sometimes), fibrous. **Flesh** whitish, pink at top of stem. **Odor** faint, subspermatic. **Spores** brown, 8–10 × 5–6 μm, rectangular with a few nodules; cystidia fusoid.

Ecology and Distribution One collection with dwarf and shrub willows at 3,355 m in Colorado. Also, with dwarf willows *S. herbacea, S. reticulata*, and *Dryas octopetala* in the alpine and with birch mixed with willows in boreal Fennoscandia. August.

Notes The dark brown cap appears smooth when the velipellis is present but is woolly underneath. It can look similar to *I. purpureobadia,* which has a smoother cap.

- Cap brown with a silvery velipellis; woolly underneath
- Stem pale brown with pink tint, fibrous
- Odor faintly spermatic
- Spores rectangular with nodules
- With willows

Reference Cripps et al. 2020. EL230–17–3, GenBank MK153678 (type); CLC 1226.

Inocybe paragiacomi E. Larss., C.L. Cripps & Vauras

Cap 1.5–4 cm across, conic-convex, with low umbo, brown, with roughened fibrils, slight velipellis. **Gills** adnexed, pale orange, old gold, well separated. **Cortina** as a few fibrils. **Stem** stout, 2–3 × 0.3–0.5 cm, equal, brownish with silvery overlay, not pruinose. **Flesh** cream, not pinkish in stem. **Odor** absent. **Spores** brown, 9–12 × 6–7.5 µm, rectangular, slightly nodulose; cystidia sub-fusiform with pedicel.

Ecology and Distribution Low alpine, in shrub willows with krummholz nearby, on rocky slope; North America (Montana) and Europe (Sweden, type). August.

Notes In contrast to *I. giacomi* and *I. subgiacomi*, the lamellae are pale orange instead of gray, the odor is not spermatic, and there are no pink tints in the stem.

- Cap conic-convex, brown, scurfy
- Gills pale orange
- Stem brownish with silvery overlay, not pruinose
- Odor absent
- Spores somewhat square with low bumps
- With willows

Reference Cripps et al. 2020. CLC 3117, Genbank MK153672.

Inocybe purpureobadia Esteve-Rav. & A. Caball.

Cap tiny, 0.5–2 cm wide, conic-convex with small umbo, umber brown to purple-brown with yellowish gray velipellis, appearing smooth but roughened, fibrous. **Gills** emarginate, well separated, cream, pale brown, yellow-brown. **Cortina** ephemeral. **Stem** 0.5–2 × 0.1–0.3 cm, equal, pale brown with pink tint, with white fibrils, floccose-pruinose at very apex. **Flesh** white, slightly pink in stem. **Odor** not distinct. **Spores** brown, 9–11.5 × 5.5–6.5 µm, slightly angular or angular-subnodulose; cystidia cylindrical-subfusiform.

Ecology and Distribution Scattered on soil between rocks on slopes in alpine zone with dwarf willow *Salix reticulata* and shrub willows in Colorado and Montana. In Finland and Spain with pine or oak, in sandy soils. August and October.

Notes This very tiny, dark brown *Inocybe* looks like a small *Telemonia*. In Europe it has been found in boreal and other non-alpine habitats. Easily overlooked.

- Cap tiny, umber brown, purple-brown
- Gills pale brown or yellow-brown
- Stem pale brown with pink tint and in flesh
- Odor not distinct
- Spores somewhat angular
- With dwarf willows

Reference Cripps et al. 2020. CLC 3109, Genbank MK153687.

Inocybe subgiacomi C.L. Cripps, Vauras & E. Larss.

Cap 1–3 cm wide, conic-convex, not umbonate, light to dark brown, can be blackish brown in center, radially fibrillose. **Gills** submarginate, well spaced, bowed out, whitish, gray-brown, orange-brown. **Cortina** not observed. **Stem** 1.5–3.5 × 0.2–0.6 cm, gradually larger toward base, whitish, pale brown, some with pink tint, pruinose at apex. **Flesh** white, with some pink in stem. **Odor** spermatic. **Spores** brown, 10–11.5 × 6–6.5 μm, only slightly nodulose; cystidia subfusiform with pedicel.

Ecology and Distribution In alpine areas of Colorado, Montana, and Wyoming with dwarf and shrub willows; also in Northern Europe. August.

Notes Similar to *I. giacomi*, but the cap of *I. subgiacomi* is typically not as tall, the cap margin is not as persistently incurved, and the odor is more strongly spermatic. *I. paragiacomi* lacks reddish tints in the stipe.

- Cap conic-convex, light to dark brown
- Gills grayish to orange-brown
- Stem pale brown with tint of pink
- Odor spermatic; spores slightly nodulose
- With willows

Reference Cripps et al. 2020. CLC 3113, Genbank MK153669.

(left) Flat cap of a *Mallocybe*, (right) Conic cap of *Pseudosperma*.

Inosperma, Mallocybe, Pseudosperma

Inosperma, Mallocybe, and *Pseudosperma* were once included in genus *Inocybe* (fiber heads). All three genera are defined by a lack of pleuro-cystidia (in contrast to *Inocybe*) and smooth brown spores—in other words, microscopic details. But it is possible to recognize them in the field. *Mallocybe* caps, gills, and stems are ocher, dark ocher, olive ocher, or reddish brown, and caps are often flat. Caps and stems can be smooth, scaly, or covered with a whitish veil. Gills are attached. The flesh is not fragile, and species lack typical *Inocybe* (spermatic) odors, but some smell like burnt sugar. Basidia contain dark pigment. They do have thin-walled cheilocystidia.

Pseudosperma species typically have a conic cap that is often yellow, buff, or brown, smooth, and rimose (splitting). Gills are almost free, often with a yellow tint; stems are smooth; the odor is usually spermatic. Caps of *Inosperma* species are convex, smooth or scaly, and odors are unusual: fishy, flowery, or sweet. Both genera lack dark pigment in their basidia; they do have cheilocystidia.

Key to Alpine Inosperma, Mallocybe, Pseudosperma

See above (pg 144) for macroscopic description of each genus. All species lack pleurocystidia. If pleurocystidia are present, see *Inocybe* (pg 142).

1. Basidia with necropigment; odor of burnt sugar, fungoid, indistinct **Mallocybe 2**
1. Basidia without necropigment; odor fishy, spermatic, flowery **Pseudosperma & Inosperma 7**

MALLOCYBE

2. Cap rather smooth and/or covered with white tissue 3
2. Cap rough, woolly, or finely scaly 5

3. Cap fibrous, smooth, ocher; odor of honey or burnt sugar **Mallocybe fibrillosa (M. dulcamara)**
3. Fruit body covered with white tissue when young; odor not distinct 4

4. Cap robust (2.5–6 cm), pale brown; stem robust **Mallocybe leucoblema**
4. Cap smaller (1.5–4 cm), pale brown, yellow-brown; stem slim **Mallocybe leucoloma**

5. Stem smooth; cap bell-shaped, dark ocher, velvety **Mallocybe arthrocystis**
5. Stem scaly; cap usually flatter, scaly or tomentose 6

6. Cap golden brown, red-brown, appressed tomentose; ring not distinct; stipe base yellow **Mallocybe fulvipes group**
6. Cap olive yellow, appressed scaly; ring zone distinct **Mallocybe terrigena** (see *M. fulvipes*)

PSEUDOSPERMA AND INOSPERMA

7. Cap brown, scaly; stem base +/-bluish; odor fishy **Inosperma calamistratum**
7. Not as above; odor spermatic 8

8. Cap conic, yellow, fibrous; stem white, equal **Pseudosperma cf. flavellum** & others
8. Cap convex, very pale brown, fibrous; stem base bulbous **Pseudosperma bulbosissimum**

Inosperma calamistratum (Fr.) Matheny & Esteve-Rav.

Cap 0.6–1.6 cm across, convex, dark brown, entirely minutely scaly. **Gills** adnate, cream then pale brown. **Cortina** buff. **Stem** 2.0–5 × 0.2–0.4 cm, equal, undulating, brown in top, bluish green (turquoise) in lower half, scaly. **Flesh** buff or reddish, bluish in stem base. **Odor** fishy. **Spores** brown, 10–14 × 5.5–7 µm, smooth, long ellipsoid; cheilocystidia clavate.

Ecology and Distribution In alpine and Arctic habitats with dwarf and shrub willows in the Rocky Mountains and Alaska. Reported from Svalbard and other Arctic-alpine habitats. Better known from subalpine forests of North America and Europe, where it is rather widespread, but not common. August.

Notes A rather distinct species with a scaly brown cap, bluish stem base, and a fishy odor. Compare to *H. subhirsutum,* which is more alpine, with wider spores, and an odor of geraniums.

- Cap convex, brown, scaly
- Gills pale brown
- Stem bluish in lower part, scaly
- Odor fishy; spores smooth
- With willows

Reference Jamoni 2008. CLC 3768, Genbank PQ191258.

Mallocybe arthrocystis (Kühner) Matheny & Esteve-Rav.

Cap 1–3 cm across, bell-shaped, ocher, dark ocher, velvety; margin clean.
Gills narrowly attached, white at first, then cream to yellow brown. **Cortina**
not noted, or as just a few fibrils. **Stem** 1–3 × 0.2–0.5 cm, equal, bit curved,
whitish then ocher, surface rough. **Flesh** whitish, rubbery. **Odor** absent.
Spores brown, 9.5–13 × 4.5–6.0 μm, smooth, long and narrow; cheilocystidia clavate, in chains.

Ecology and Distribution Occurring in alpine areas of Colorado, Montana,
and Alaska. Known from Arctic-alpine areas in Europe and Greenland. Also
found at lower elevations. With willows in August in the alpine.

Notes The bell-shaped cap is distinct from most other Mallocybes, as are
the strikingly white gills of young specimens. _M. fibrillosa_ (_I. dulcamara_)
caps are smoother, and _I. fulvipes_ caps are darker. The long, narrow spores
are distinctive.

- Cap bell-shaped, ocher, velvety
- Gills white to yellow-brown
- Stem white to ocher, rough
- Spores long and narrow, smooth
- With dwarf and shrub willows

Reference Cripps, Larsson, Horak 2010. CLC 1293. CLC 1326, Genbank
GU980649.

Mallocybe fibrillosa (Peck) Matheny & Esteve-Rav.

Cap 1.5–4.5 cm across, convex becoming flat, sometimes with umbo, ocher (dull yellow-brown), smooth to fibrous. **Gills** adnate, pale ocher, ocher, some with olive tint. **Cortina** whitish buff. **Stem** 1.5–4.0 × 0.3–0.6 cm, equal, pale ocher, often whiter at base, rough-fibrous. **Flesh** white in cap, watery yellow in stem, which is tough. **Odor** of burnt sugar, honey-like. **Spores** brown, 8.5–11.5 × 5–6 µm, smooth, slightly bean-shaped; cheilocystidia clavate.

Ecology and Distribution In North America, occurring in the Rocky Mountain alpine, Alaska, and Western subalpine forests. In the alpine, found with dwarf and shrub willows, and possibly *Dryas*. Known from Arctic, alpine, and boreal areas of the Alps and Fennoscandia with willow and birch. August.

Notes Previously, this *Mallocybe* was called *Inocybe dulcamara* (=*Mallocybe dulcamara*). Several alpine forms have been described under this name. The species has a rather smoothish cap compared to similar Mallocybes in the alpine.

- Cap ocher, smoothish, flat in center when mature
- Gills light ocher; cortina whitish
- Stem light ocher, tough, white at base
- Odor of burnt sugar; spores smooth
- With willows, *Dryas*, conifers, and broadleaf trees

References Cripps, Larsson, Horak 2010; Matheny et al. 2023. CLC 1241, Genbank GU980637. CLC 1295, Genbank GU980636.

Mallocybe fulvipes (Kühner) Matheny & Esteve-Rav. group

Cap 2–5 cm across, broadly convex to rather flat, orange-brown, red-brown, fibrillose-scaly. **Gills** adnate, pale golden yellow, orange-brown. **Cortina** whitish yellow. **Stem** robust, 1.5–4.0 × 0.5–1 cm, equal, yellow-brown, scaly. **Flesh** bright yellow in stem base; rubbery. **Odor** faint. **Spores** 9–12 × 4.5–5.5 µm, smooth, ellipsoid to slightly bean-shaped; cheilocystidia cylindrical.

Ecology and Distribution In the Rocky Mountain alpine zone, from Montana to Colorado. Known from the Alps, Norway, and Sweden. With willows in August.

Notes This may be a complex that includes the similar *Mallocybe substraminipes*. One of the darkest Mallocybes, *M. fulvipes* can be recognized by the bright yellow tissue inside the stem base. *Mallocybe terrigena*, better known from subalpine forests, can occur in the alpine and has yellower colors and a distinct ring.

- Cap orange-brown, red-brown, fibrillose-scaly
- Gills pale golden
- Stem yellow-brown, scaly; stem yellow inside
- Odor not distinct; spores smooth
- With willows

Reference Cripps, Larsson, Horak 2010. CLC 3534 above; CLC 2292, Genbank GU980604.

Mallocybe leucoblema (Kühner) Matheny & Esteve-Rav.

Cap robust, fleshy, 2.5–6 cm across, convex, dull brown, fibrous-smooth, covered with white tissue when young. **Gills** attached to subdecurrent, pale olive brown, well spaced. **Cortina** white. **Stem** robust, 2.5–5.0 × 0.8–1.2 cm, equal, pale brown, olive brown, covered with white tissue at first. **Flesh** pale yellow, rubbery. **Odor** faint. **Spores** 9–11 × 5–6 µm, smooth, slightly bean-shaped; cheilocystidia clavate.

Ecology and Distribution In alpine habitats with willows and *Dryas* in the Rocky Mountains and also in subalpine forests. Reported from alpine, boreal, and subalpine habitats in Europe. August.

Notes This *Mallocybe* is recognized by its robust stature and white covering. The caps and stems of *M. leucoloma* are also covered with white tissue, but this species is much slimmer, less robust.

- Cap robust, fleshy, brownish, at first covered with white tissue
- Gills pale olive brown; can be subdecurrent
- Stem robust, covered with white tissue
- Odor not distinct; spores smooth
- With willows and *Dryas*

Reference Cripps, Larsson & Horak 2010. CLC 3906; CLC 1721, Genbank GU980632. Above EL102–09, Sweden.

Mallocybe leucoloma (Kühner) Matheny & Esteve-Rav.

Cap 1.5–4 cm across, hemispherical then conic convex to almost flat, pale yellow-brown, orange-brown, covered with white tissue at first or on cap margin, rough fibrous. **Gills** decurrent to adnate, olive-yellow, mustard color. **Cortina** white, copious. **Stem** 1.5–4.5 × 0.2–0.6 cm, equal, pale brown, covered with white tissue. **Flesh** pale yellow-brown. **Odor** faint, fungoid. **Spores** 9–11 × 4.5–6 µm, smooth, almost ellipsoid; cystidia clavate.

Ecology and Distribution Reported from many places in the Colorado alpine with shrub and dwarf willows. Also from alpine areas of Alaska and Europe. August.

Notes This species is much slimmer than *Mallocybe leucoblema,* which also has a white covering. Older specimens can look like *M. fibrillosa.* DNA sequences of Rocky Mountain specimens match the type described by Kühner from the Alps.

- Cap conic-convex to almost flat, dark ocher, rough
- Gills olive to mustard yellow
- Stem and cap covered with white tissue at first; slim
- Odor not distinct; spores smooth
- With willows

Reference Cripps, Larsson, Horak 2010. NS6085 above. CLC 1431, Genbank GU980619.

Pseudosperma bulbosissimum (Kühner) Matheny & Est.-Rav.

Cap 1–3.5 cm, conic-convex, with small umbo or not, pale yellow, pale gold, radially fibrous; margin with white tissue. **Gills** almost free, narrow, white, pale to dark yellow; edges white. **Cortina** as a few fibrils. **Stem** 1.5–6 × 0.4–0.6 cm, gradually enlarging, with a small bulb at base, white, fibrous. **Flesh** white with yellow tint. **Odor** faintly spermatic. **Spores** 11–13 × 6–8 μm, smooth, slightly bean-shaped; cheilocystidia almost cylindrical.

Ecology and Distribution Occurring in the Colorado alpine with dwarf willows *Salix arctica* and *S. reticulata*. Also known from the French and Italian Alps with dwarf willows *S. herbacea* and *S. reticulata*. August.

Notes Distinguished from other whitish or pale capped *Inocybe* or *Pseudosperma* species by the onion bulb at the stem base. However, the bulb is not always obvious. Yellow on the gills suggests it might be a *Pseudosperma*.

- Cap conic-convex, pale yellowish, rimose
- Gills yellowish
- Stem white, with a bulbous base
- Odor faintly spermatic or not; spores smooth
- With willows

References Bon 1992; Ferrari 2006. CLC 1701, Genbank PQ191446.

Pseudosperma cf. *flavellum* P. Karst.

Cap 2–3.5 cm across, tall conic-convex with rounded apex, yellow brown, radially fibrous; margin rimose, splitting. **Gills** narrowly attached, golden yellow, staining brown; edges lighter. **Cortina** as a few fibrils. **Stem** 2.5–4.5 × 0.3–0.4 cm, gradually enlarging toward base, cream, striate. **Flesh** yellow in cap, cream in stem. **Odor** faintly spermatic. **Spores** 9–11 × 5–6 µm, smooth, somewhat bean-shaped; fat yellow basidia present; cheilocystidia clavate.

Ecology and Distribution With willows in the Rocky Mountain alpine, at least in Montana and Wyoming and likely more widespread. Also known from alpine areas of Sweden and Finland. August.

Notes The ITS sequence of this member of the *P. rimosum* group matches that of E. Larsson's *P.* cf. *flavellum* (sequence in Genbank) to be described as a new alpine species.

- Cap tall, conic-convex, golden yellow
- Gills almost free, and yellow
- Stem white, fibrous
- Odor faintly spermatic; spores smooth
- With willows

Reference Genbank match to MH310766.1. CLC 3087, Genbank PQ191453.

Laccaria in Alpine and Arctic Habitats

Although numerous species of *Laccaria* have been reported from Arctic and alpine habitats in the Northern Hemisphere, we found only five during twenty years of collecting in the Rocky Mountains. More potentially exist, especially in the North American Arctic. The genus *Laccaria* is relatively easy to identify because of the overall bright to pale orange colors with pinkish tones of the whole fruiting body. While some species can have lilac colors, none were encountered in the Rocky Mountain alpine. Caps are often greasy, and can be striate or smooth, translucent or opaque (especially on drying). Gills are pink or orange, thick and waxy, and attached to the stem. Stems vary from smooth to fibrillose and are usually the same color as the cap. Some species are very tiny in the alpine and easily overlooked. The white spores are spiny, which is a confirming character. A difficulty for field keying is that a knowledge of the size and shape of the spores, plus the number per basidium, is often required for identifying to species. Counting the number of gills can help.

Hygrocybe species can be orange and also have thick, waxy gills, but spores are white and smooth. *Loreleia* species have dark orange caps but occur on liverworts; and *Lactarius lanceolatus* is orange but is larger, with amyloid warted spores. A few Telemonias have orange greasy pointed caps but brown spores. Entolomas can have pink gills but pink angular spores.

Top: *Laccaria laccata* var. *pallidifolia* CLC 1238b. Bottom: Alpine *Laccaria* can be small and can fade on drying.

Laccaria species are mycorrhizal primarily with willows and birch in alpine and Arctic habitats and with conifers in the krummholz zone, subalpine forests, and other low-elevation habitats.

Key to Laccaria in Alpine and Arctic Habitats

Cap and stem orange, reddish brown to buff; gills pinkish in all species included here.

1. Cap 0.5–2.5 cm; ≤ 24 full gills; stem 1–3 mm wide; spores 9–14 µm **2**
1. Cap 2–5 (–7) cm, ≥ 24 full gills; stem 2–8 mm wide; spores 7–10 µm **4**

2. Basidia 2 spored; gills attached to decurrent **Laccaria pumila**

2. Basidia 4 spored; gills usually attached 3

3. Fruiting body orange to orange-red, bit larger **Laccaria montana**

3. Fruiting body more red-brown, smaller; so far only from 3 locations in Colorado **Laccaria pseudomontana**

4. Stem smooth to fibrous; cap 1–2(–3) cm, striate **Laccaria laccata var. pallidifolia**

4. Stem striate, rough-fibrous; cap 2–5(–7) cm, not striate **Laccaria nobilis**

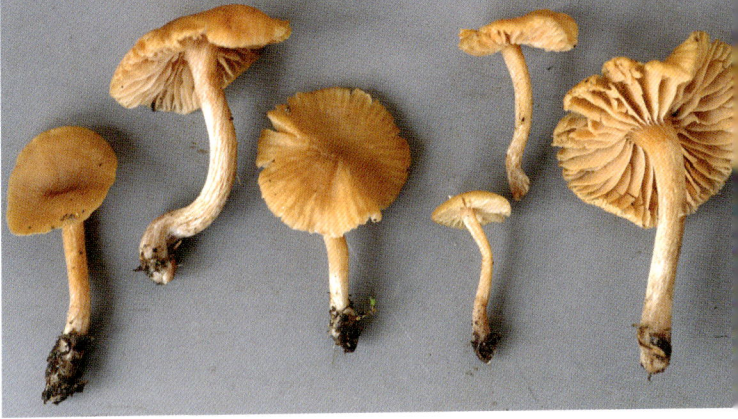

Laccaria laccata* var. *pallidifolia (Peck) Peck alpine form

Cap 2–3.5 cm wide, convex, and then flat or depressed in center, indistinctly striate, pale orange, dark orange, drying lighter, smooth, sometimes greasy. **Gills** adnate to subdecurrent, thick, well spaced, pale orange or pink; more than 24 reach the stem. **Stem** 1.5–5 × 0.2–0.5 cm, long and thin, equal or gradually enlarging at base, pale orange, smooth to minutely fibrillose, finely striate. **Flesh** white or pale orange. **Odor** not distinct. **Spores** 7–9 × 6–8.5 μm, subglobose, ellipsoid, with short spines, 4 per basidium.

Ecology and Distribution Rare in the Colorado alpine, with dwarf willow *S. reticulata,* shrub willow *S. glauca, Dryas octopetala,* and birch. Widely reported in other Arctic and alpine habitats, although reports need verification. More common in a wide array of habitats below treeline with many different hosts. August.

Notes *Laccaria laccata* is typically slightly larger and has more gills than either *L. pumila* or *L. montana.* The cap is striate and the stem smoother than for *L. nobilis.*

- Cap orange, striate, greasy
- Gills pink-orange, thickish, well separated, ≥ 24
- Stem pale orange, smooth or with few fibrils
- Spores 7–9 × 6–8.5 μm, roundish, spiny, 4 per basidium
- With willows, *Dryas*, and possibly birch

Reference Osmundson, Cripps, Mueller 2005. CLC 1655, Genbank DQ149847.

Laccaria montana Singer

Cap 0.5–2.5(–3) cm wide, convex, then flat, sometimes depressed in center, smooth to minutely scaly, striate, orange-brown, red-brown, brick red, drying to pale orange-buff. **Gills** adnate, rarely short decurrent, thick, well separated, somewhat broad, orange, pinkish orange. **Stem** 1–3 × 0.2–0.4 cm, equal, orange-brown or red-brown, same as cap, smooth or with few fibrils. **Flesh** white to pale orange. **Odor** not distinct. **Spores** 8–11 × 8–10 µm, almost round, with short spines, 4 per basidium.

Ecology and Distribution In the alpine with shrub willow and birch, often in moss. Possibly more common in the northern rather than the southern Rocky Mountains. Known from Arctic and alpine habitats in Europe and Svalbard with willows, and more widely distributed in lower-elevation habitats. July-August in the alpine.

Notes Similar to *Laccaria pumila* and *Laccaria tortilis* in AA habitats, but these species have 2-spored basidia. Small forms of *Laccaria laccata* can look similar but have smaller spores. *Laccaria pseudomontana* is darker red and is known from only three places in Colorado. August.

- Cap orange, striate, greasy when fresh, paler dry
- Gills pink-orange, thick, well separated, ≤ 24
- Stem reddish orange, orange, rather smooth
- Spores 8–11 × 8–10 µm, round, spiny, 4 per basidium
- With willows and possibly birch

Reference Osmundson, Cripps, Mueller 2005. CLC 4113 above; TWO 319. Genbank DQ149862.

Laccaria nobilis A.H. Sm. apud G.M. Mueller

Cap 2–5(–7) cm across, convex, some with shallow central depression, pale pinkish orange to dark orange, drying lighter, smooth when young, becoming minutely scaly, not striate. **Gills** adnate to adnexed, broad, thick, well spaced, pink. **Stem** 2–5 × 0.3–0.8 cm, robust, equal or gradually enlarged toward base, pale whitish orange or orange-brown, tough, striate, rough-fibrous; at base white to violet. **Flesh** white, pale orange, or violet. **Odor** not distinct. **Spores** 6–8 × 5–6 μm, ellipsoid, with short spines, 4 per basidium.

Ecology and Distribution In the low alpine in Colorado with mixed dwarf and shrub willows. This western North American species is typically reported at lower elevations with conifers. Rare in AA habitats. August.

Notes This rather robust species is unusual for alpine and Arctic habitats but is clearly with willows. *Laccaria bicolor* is also robust and sometimes has violet basal mycelium, and it is best separated molecularly, but so far it has not been reported from AA habitats.

- Cap robust, orange, smooth to minutely scaly, not striate
- Gills pink, thick, well separated, ≥ 24
- Stem robust, orange, rough-fibrous striate
- Spores 6–8 × 5–6 μm, ellipsoid, spiny, 4 per basidium
- With willows

Reference Osmundson, Cripps, Mueller 2005. CLC 1304, Genbank DQ149859.

Laccaria pseudomontana Osmundson, C.L. Cripps, G.M. Mueller

Cap 0.4–1.8 cm, convex, later flat, dark orange to red-brown, smooth, slightly striate. **Gills** adnate to subdecurrent, thick, well separated, ≤ 24, pink. **Stem** 1–1.5 × 0.1–0.2 cm, equal, smooth to fibrous, dark orange-red or red-brown. **Flesh** white. **Odor** not distinct. **Spores** 8–10 × 7–8.5 µm, almost round, with short spines, 4 per basidium.

Ecology and Distribution Reported only from the Colorado alpine in the 10-mile Range near Blue Lake Dam and the San Juan Mountains. Among mosses, near shrub willows *Salix glauca* or *S. planifolia* mixed with birch. August.

Notes Most similar to *Laccaria montana*, it is best distinguished by DNA comparison, although in the field, fruiting bodies of *Laccaria pseudomontana* appear to be smaller and darker red. Spines on the spores are somewhat smaller than for other alpine species.

- Cap small, dark orange-red, red-brown, smooth
- Gills pink, thick, well separated, ≤ 24
- Stem dark red, smooth to fibrous
- Spores 8–10 × 7–8.5 µm, spiny, rather round
- With willows, possibly birch

Reference Osmundson, Cripps, Mueller 2005. CLC 1682 above. CLC 1625 Type, Genbank DQ149871.

Laccaria pumila Fayod

Cap 0.5–1.5 cm wide, convex to flat or even depressed in center, pale orange to nearly red-brown, paler on drying, striate, smooth or minutely fibrillose, greasy or not. **Gills** adnate to short decurrent, somewhat thick, well separated, ≤ 24, pale orange, pinkish orange. **Stem** 1.5–3.5 × 0.1–0.4 cm, equal, pale pinkish orange, orange, dark red-brown, smooth or minutely fibrillose. **Context** white to pale orange. **Odor** not distinct. **Spores** 9–12 × 8–10 μm, almost round, with short spines, 2 per basidium.

Ecology and Distribution In Arctic and alpine habitats, usually among mosses primarily with shrub willows. Also in AA habitats in Europe and Svalbard, and occasionally in other habitats. August.

Notes In Arctic-alpine habitats, *Laccaria tortilis* also has 2 spores per basidium, but the spores have longer spines. *Laccaria pumila* is similar to *Laccaria montana,* which has 4 spores per basidium. *Laccaria laccata* is usually larger, with smaller spores and more gills.

- Cap orange, striate, greasy fresh, slightly striate
- Gills pink-orange, thick, well separated, ≤ 24
- Stem reddish orange, smooth, or with few fibrils
- Spores 9–12 × 8–10 μm, round, spiny, 2 per basidium
- With willows

Reference Osmundson, Cripps, Mueller 2005. CLC 2863 above; CLC 1252, Genbank DQ149864.

Lactarius lanceolatus.

Lactarius in Arctic and Alpine Habitats

In other habitats, *Lactarius* species are recognized by a dished or concave cap, decurrent gills that ooze a latex (milk) when cut, and a light-colored spore print. However, these features are not definitive in cold-climate species. First, caps can be extremely small, with several fitting on the head of a penny! The caps often appear flat and not dished. Because these nanoforms are short, the gills are not always clearly decurrent. To further complicate identification, cold, dry conditions are not conducive to latex production. There is not much to taste, so it can be difficult to tell whether the flesh is hot, which is important to know. A couple of clues that might make you suspect a *Lactarius* are that some of the smaller species have a greasy cap, pale salmon-colored gills, and an odor reminiscent of cocoa butter. In some, gills will turn color when cut, especially lavender, even if they don't exude milk. The spores are white. With a microscope and Melzer's solution, the amyloid warts characteristic of the family and genus can be observed. None are good edibles in Arctic-alpine habitats.

Quite a diversity of Lactarii is known from Arctic-alpine habitats, and they are mycorrhizal with willows or birch. Most of our species

Lactarius nanus.

appear to have circumpolar distributions and occur in Greenland, Iceland, Svalbard, and AA habitats in Europe and Asia, as well as in North America. Our one exception, *L. repraesentaneus,* usually a subalpine species, is reported here with alpine willows.

Key to Alpine and Arctic Lactarius

1. Cap, gills, and stem bright orange **Lactarius lanceolatus** (other AA orange species exist outside the Rocky Mountains)
1. Not as above 2

2. Gills staining violet when cut or bruised 3
2. Not as above, but gills can bruise gray 6

3. Cap 6–10 cm, robust, yellow-brown; margin shaggy **Lactarius repraesentaneus**
3. Not as above, cap smaller, 2–5 cm 4

4. Cap brownish lavender, slightly zoned **Lactarius brunneoviolaceus**
4. Cap another color, not zoned 5

5. Cap, gills, and stem yellow-cream **Lactarius salicis-reticulatae**

5. Cap brownish white; stem white **Lactarius pallidomarginatus**

6. Odor of coconut; cap pinkish buff, pale gray-brown; with birch **Lactarius glyciosmus**

6. Odor not distinct; cap uniform brown; with willows **Lactarius nanus**

Lactarius* aff. *brunneoviolaceus M.P. Christ.

Cap 3–6 cm across, convex, depressed in center, smooth, greasy, slightly zonate, brown with violet tints, marbled-looking, whitish toward margin. **Gills** attached, cream, staining violet when bruised. **Stem** 2.5–3.0 × 2.0–2.5 cm, clavate, smooth, white, staining violet where bruised. **Flesh** cream. **Milk** white, staining tissue violet. **Odor** sweet. **Taste** mild to bitter. **Spores** white, 8–12 × 6.5–8 μm, ellipsoid, with amyloid ridges.

Ecology and Distribution Our collection from the Beartooth Plateau was in the krummholz zone near both spruce and willow. Also reported from Arctic-alpine areas including Finland, Iceland, and Svalbard near willows. August.

Notes We have one sighting of this species, and more are needed to determine its habitat and range. The cap stains green in KOH. *Lactarius montanus* is similar and occurs at treeline but is more uniformly brown; the cap also turns green in KOH. The violet-staining *Lactarius pallidomarginatus* is smaller and slimmer.

- Cap robust, marbled brownish violet, slightly zoned
- Gills white, staining violet, with white milk
- Odor sweet, taste mild
- Spores white with amyloid warts
- With willows and conifers in krummholz

References Barge and Cripps 2016a, b. CLC 3098, Genbank KX394283.

Lactarius glyciosmus (Fr.) Fr. alpine form

Cap 1.5–5 cm across in alpine, convex to flat, sometimes depressed, often with small papilla, smooth, dry, not zoned, pale grayish brown, pale mauve-gray, or pale salmon with a hoary (frosty) coating when young. **Gills** adnate to decurrent, cream, pinkish buff, or pale yellowish orange. **Stem** 1–4 cm long up to 1 cm wide, slightly clavate, pale salmon with hoary coating. **Flesh** cream. **Milk** scarce, watery, white, unchanging. **Odor** of coconut. **Taste** mild to slightly acrid. **Spores** white, 7–9 × 5–7 μm, ellipsoid, with amyloid warts and ridges.

Ecology and Distribution This alpine form occurs in North America with bog birch, fruiting in late summer to fall, usually August. The species in general is widespread in the Northern Hemisphere in temperate, boreal, and Arctic-alpine areas, where it is mycorrhizal with birch.

Notes The key to recognizing this species is the odor of coconut or cocoa butter (Coppertone) and the presence of low bog birch. In Montana, the tiny alpine form occurs in the few areas where bog birch exists; the larger subalpine form occurs with birch trees below treeline.

- Small, mauve to salmon frosty cap
- Pinkish buff to yellow-orange gills; milk watery white
- Odor of coconut
- Spores white with amyloid warts
- With birch

References Barge and Cripps 2016a, 2016b. CLC 1380; CLC 1624, Genbank KR090907.

Lactarius lanceolatus O.K. Mill. & Laursen

Cap 1.0–4.5 cm across, convex, then depressed, often with a small papilla, smooth, sometimes faintly scaly in center, viscid to dry, not zoned, deep orange-brown, deep orange; margin curved under, slightly wavy. **Gills** adnate to subdecurrent, cream, pale yellow, pale orange, discoloring brownish. **Stem** 1.0–2.0 × 1 cm wide, equal to clavate, smooth, dry, pale orange with hoary coating, hollow. **Flesh** pale orange. **Latex** scarce, watery, white, unchanging. **Odor** mild. **Taste** mild. **Spores** white, spores 8–10 × 6–8 μm, ellipsoid with amyloid warts.

Ecology and Distribution In the Northern Rockies with dwarf willows *Salix reticulata* and *S. arctica*, and shrub willow *S. planifolia,* fruiting in late summer. Also in Alaska. Occasional in other Arctic-alpine areas in the Northern Hemisphere; mycorrhizal with willows.

Notes This is the only orange *Lactarius* known so far above treeline with willows in the Rocky Mountains. The species was described from Alaska, and DNA of Montana specimens match exactly. Other potential orange species from European and other Arctic-alpine habitats include *Lactarius aurantiacus* and *Lactarius lapponicus.* In krummholz areas, orange subalpine species must be considered.

- Cap, gills, and stem orange
- Latex white, watery
- Taste mild
- Spores white with amyloid warts
- With willows

References Barge and Cripps 2016a, 2016b. CLC 3103c above; CLC 2358, Genbank KR090918.

Lactarius nanus J. Favre

Cap 1–5 cm across, shallowly convex, with sunken center, greasy to dry, medium to dark brown with hoary (frosty) coating; margin turned down. **Gills** adnate to decurrent, cream to pale apricot, discoloring grayish where bruised. **Stem** 0.5–3.0 cm × 0.3–1.5 cm, equal, smooth, pale apricot with hoary coating. **Flesh** cream. **Odor** indistinct. **Milk** scarce, watery, white, unchanging. **Taste** mild to slightly acrid. **Spores** white, 7–10 × 5–8 µm, ellipsoid, with amyloid warts.

Ecology and Distribution Uncommon or difficult to find in the Rocky Mountains. But widespread in many Arctic-alpine areas in the Northern Hemisphere; mycorrhizal with dwarf willows such as *S. arctica* and *S. reticulata*. Late summer to fall.

Notes *Lactarius nanus* can be so small that it is difficult to recognize as a *Lactarius*. The gills are barely decurrent, and it rarely oozes milk when cut. The brown cap and apricot-colored gills are the best features for recognition. It does not stain lavender when cut, as some other alpine Lactarii do. The similar *Lactarius glyciosmus* is with birch and has a coconut odor.

- Cap small, brown, greasy
- Stem and gills salmon color
- Odor not distinct; milk scant, watery white
- No lavender staining
- Spores white with amyloid warts
- With dwarf willows

References Barge and Cripps 2016a,b. CLC 3118 above; CLC 1403 Genbank KR090927.

Lactarius pallidomarginatus Barge & C.L. Cripps

Cap 2–5 cm across, convex, or slightly sunken in center, blotchy brown, grayish brown, with violet tints and pale margin, smooth, dry to slightly greasy. **Gills** subdecurrent, white, cream, staining violet when damaged. **Milk** scarce, watery, staining tissue violet. **Stem** 1–4 × 0.5–1.0 µm, equal, white, staining violet when damaged, dry. **Flesh** white, staining violet. **Odor** mild. **Flavor** mild. **Spores** white, 8–10 × 6.6–8 µm, ellipsoid, with amyloid warts and thin ridges.

Ecology and Distribution Mycorrhizal with willows in the southern Rocky Mountains. Appears to be endemic to this region. August.

Notes Similar in cap color and violet staining reactions to *Lactarius Pseudouvidus* which is reported in the European alpine, *L. montanus* in the Rocky Mountain krummholz zone, and *L. brunneoviolaceus* in both habitats.

- Cap blotchy brown with pale margin
- Gills and stem white to cream
- Cap, gills, and stem staining violet when damaged
- Spores white with amyloid warts
- With willows

References Barge and Cripps 2016a,b. EB 0041, Genbank KR090940.

Lactarius repraesentaneus Britzelm.

Cap robust, 6–10 cm across, convex with a depressed center, smooth at center, hairy at margin, viscid to dry, golden yellow. **Gills** adnate to decurrent, crowded, cream to pale yellow, staining violet. **Stem** 3–7 × 1.5–3.5 cm, stout, equal to slightly clavate, viscid to dry, pitted, cream to pale golden brown, hollow. **Flesh** white, staining violet. **Milk** white, becoming violet. **Odor** fruity, spicy-floral. **Taste** somewhat hot to very hot. **Spores** white, 8–10.5 × 6–8.5 µm, ellipsoid, with amyloid ridges.

Ecology and Distribution In the alpine krummholz zone on the Beartooth Plateau, occasionally with pure willow, but often with willow mixed with conifers. Widespread in the Northern Hemisphere in temperate, boreal, and subalpine habitats near conifers. August.

Notes Typically a subalpine species with conifers in the Rocky Mountains, this *Lactarius* has invaded the alpine in some areas, including Quebec. Its distinctive features and robust stature make it easily recognizable. While meaty, the acrid taste precludes it being a good edible.

- Cap fleshy, golden with a shaggy margin
- Gills staining violet when damaged
- Odor fruity, taste hot
- Spores white with amyloid warts
- Krummholz zone, near willows or conifers

References Barge and Cripps 2016a,b; Hutchinson et al. 1988. CLC 2318, Genbank KR090948.

Lactarius salicis-reticulatae Kühner

Cap 2–4 cm across, irregularly convex, with a depressed center, greasy to dry, smooth, pale cream-yellow, staining violet when bruised; margin turned down. **Gills** adnate to subdecurrent, somewhat separated, pale yellow, pale apricot, sometimes pink tinted, staining violet when damaged. **Stem** 1.5–2.0 × 1.0–1.5 cm, equal or slightly clavate, greasy to dry, cream, pale yellow, staining violet, hollow. **Flesh** white, staining violet. **Milk** scarce, watery white, becoming violet. **Odor** absent to slightly sweet. **Taste** mild. **Spores** 8.5–11.5 × 8–10 µm, subglobose, with amyloid ridges.

Ecology and Distribution In the Rockies with dwarf willows *S. arctica* and *S. reticulata* mixed with shrub *Salix* on the Beartooth Plateau. Widespread in Arctic-alpine areas in the Northern Hemisphere with willows. August.

Notes Several violet-staining species are reported from Arctic-alpine Alaska that have not been found in the Rocky Mountains. *Lactarius aspideus* is one that is also yellow with violet staining gills, but it is not molecularly related.

- Cap and stem pale yellow
- Gills pale salmon with pinkish tint
- Gills staining violet when damaged
- Odor and taste mild
- Spores white with amyloid warts
- With willows

References Barge and Cripps 2016 a, b. CLC 2776, Genbank KR090957.

Leccinum in Finland.

Leccinum in Alpine and Arctic Habitats

"Boletes" have a fleshy cap and stem, and pores instead of gills under the cap. The pores are the openings of parallel tubes that hold the spores. While there are many genera in the Boletaceae, *Leccinum* is the only one known to occur regularly in Arctic or alpine habitats, and species occur with dwarf birch (*Betula nana, B. rotundifoliae*) and not willows. Fruitings can be prolific in these situations. Several *Leccinum* species are found in Arctic and alpine habitats, including *Leccinum scabrum, L. holopus*, and *L. variicolor*, but only *L. rotundifoliae* appears restricted to these cold climates. The others are also possible in subalpine and boreal forests. We have found only *Leccinum rotundifoliae* above treeline in the Rockies, and with dwarf birch.

Suillus species are excluded from Arctic alpine habitats due to the lack of appropriate hosts, but they may occur in the krummholz zone near conifers. A similar situation exists for *Boletus* species, although we have seen *Boletus edulis* fruiting in typical alpine vegetation with *Dryas,* which can host non-alpine species above treeline.

Leccinum rotundifoliae (Singer) A.H. Sm., Thiers & Watling

Cap 6–9 cm across, broadly convex, slightly greasy or dry; cream, pale brown, smooth, becoming finely cracked. **Pores** tiny (2/mm), whitish, staining brown. **Stem** 7–13 × 2–3 cm, enlarged toward base, whitish, covered with pale grayish brown scales. **Flesh** white, unchanging or turning slightly pinkish but not blue. **Odor** slightly sweet. **Spores** brown, 16–20 × 5–7 μm, long fusoid.

Ecology and Distribution In the Rocky Mountains and Alaska, mycorrhizal with dwarf birch *Betula nana* in alpine and Arctic habitats. Also known from Arctic and alpine areas of Europe, Svalbard, Greenland, and the Altai Mountains of Russia, always with birch. August.

Notes *Leccinum* species in AA habitats are restricted to birch, and birch is rare in the Rocky Mountain alpine. The Arctic *Leccinum scabrum* has an orange cap and is separated molecularly. *Leccinum holopus* is white and stains bluish, and is not usually alpine, but it can occur at high elevations.

- Cap cream to pale brown, smooth or cracked
- Pores whitish
- Stem whitish with grayish black scales
- Flesh white, unchanging or turning a bit pink
- With dwarf birch

Reference Den Bakker et al. 2007. CLC 2859.

Russula in Arctic-Alpine Habitats

Russulas in Arctic-alpine habitats are typically much smaller than their forest relatives. "Nano-forms" are likely an adaptation to cold climates or a result of nutrient limitations from mycorrhizal hosts, which are primarily willows, birch, and *Dryas*. *Russus* is Latin for red, and many Russulas in AA habitats do have reddish caps. Other possible colors are white, yellow, orange, maroon, and magenta. Gill and spore colors range from white to dark ocher; a spore print can be essential for identification. Stems are typically white and shaped like a piece of chalk, but some are flushed pink. Stems can be quite fragile or brittle. Odor and taste are important in identification and should be tested immediately in the field. Some species such as *R. nana* can be radish hot early in the morning but less hot later in the day. Similarly, odors fade quickly in extreme environments and should be noted in the field. Melzer's reagent on spores confirms whether they have the amyloid warts characteristic of the genus. Russulas in the Rocky Mountain alpine are but a subset of potential species in AA habitats globally. None are considered good edibles, although *R. subrubens* is a possibility; it is best left for conservation. Other red-capped mushrooms in AA habitats include Hygrocybes, which have waxy caps and gills.

Russula in dwarf willows.

Key to Alpine Russula Species

1. Flesh dense; cap whitish, depressed in center (*Lactarius*-like) **Russula laevis**
1. Flesh more fragile; cap more highly colored; cap center not depressed 2

2. Odor fishy; with willows 3
2. Odor not fishy; with willows, *Dryas,* or birch 4

3. Cap small, 2.5–3.5 cm, magenta, gills yellowish; taste slightly acrid **Russula neopascua**
3. Cap robust, 3–8 cm, mottled orange-brown; taste mild **Russula subrubens**

4. Cap robust, orange; taste hot to mild; with birch, *Dryas*, willows **Russula intermedia**
4. Cap usually smaller, more reddish, rosy, maroon, magenta, olive 5

5. With birch 6
5. With willows 7

6. Cap rosy, striate to center, thin-fleshed **Russula sphagnophila**
6. Cap mixed olive, yellow, magenta, pink, indistinctly striate at margin **Russula altaica**

7. Gills and spores to dark yellow; cap deep red (yellow-brown young) **Russula purpureofusca**
7. Gills and spores white, cream, light yellow; cap red to maroon 8

8. Cap cherry red, fading to white; slightly hot **Russula nana**
8. Cap more maroon, magenta, ruby red with dark center; mild to hot 9

9. Taste hot; stem white; cap maroon, magenta, with dark center **Russula laccata**
9. Taste mild to slightly hot; often with pink on stem; cap ruby red 10

10. Gills light yellow; taste mild **Russula saliceticola**
10. Gills white; slightly hot (Alaska) **Russula alpigenes**

Russula cf. *alpigenes* (Bon) Bon

Cap 2–6 cm across, convex to almost flat, or dished in center, deep ruby red to almost black in center, smooth, greasy; margin not striate. **Gills** adnate, narrow to broad, white, then cream, some pinkish near cap margin. **Stem** 2.5–6 × 0.8–1.8 cm, long clavate, narrower in middle, white, with pinkish stains in age, smooth. **Flesh** white, solid; pink under cap cuticle. **Odor** fruity-sweet or absent. **Taste** slightly acrid, bitter, or mild. **Spores** white, 8–9.5 × 6.5–7.5 μm, ellipsoid with amyloid warts connected by thin ridges.

Ecology and Distribution In the low alpine and at treeline; here reported from Alaska with *Dryas* mixed with willows. Also known from the French Alps. August.

Notes This species could be confused with a number of deep red species, but only *Russula nana* and *Russula laccata* also give a white spore print and have a slightly acrid taste, and they lack pink on the stem. *Russula saliceticola* may be similarly colored and the stem can also have pink tints, but it has a mild taste and slightly darker spores and gills. So far, *Russula alpigenus* has been reported only in Alaska for North America.

- Cap ruby red with dark to black center
- Gills and spores white
- Stem white, sometimes with pink tint
- Odor fruity or absent; taste slightly acrid or not
- Spores white with amyloid warts and ridges
- With willows

Reference Noffsinger 2020. CLC 3822b, Genbank MT583303.

Russula cf. *altaica* (Singer) Singer

Cap 1–4 cm across, convex, some with sunken center, mottled olive yellow, olive green, grayish magenta, darker at center to umber or lighter to pinkish, smooth, greasy; margin not or indistinctly striate; cuticle thick, gelatinous, peeling. **Gills** adnexed, crowded, thickish, white, graying somewhat. **Stem** 2–3 × 0.5–1.5 cm, clavate, white with faint or strong pink blush. **Flesh** white, slightly graying, vinaceous beneath cuticle; stipe stuffed. **Odor** indistinct or absent. **Taste** mild, possibly bitter, not acrid. **Spores** pale cream, 7.5–9 × 5.5–6.5 µm, broadly ellipsoid with isolated amyloid warts.

Ecology and Distribution Likely with bog birch at treeline in Colorado, but willows and *Dryas* were nearby. Also reported from the Altai Mountains of Russia, Svalbard, Greenland, and Canada with birch. August.

Notes Recognized by the varied pileus colors and pink-flushed stem; reminiscent of the forest species *Russula queletii*. However, young specimens can be simply dark reddish purple. This is the only species reported with alpine birch in Colorado so far. *Russula* cf. *sphagnophila* is found with birch in Alaska; it has a delicate, striate, rosy-purple cap and vinaceous tints at the top of the stipe. *Russula intermedia* is also possible with birch in some areas but is orange-brown.

- Cap a mix of magenta and olive green
- Gills white, stem flushed pink
- Taste mild, odor not distinct
- Spores cream with amyloid warts
- With birch

Reference Noffsinger and Cripps 2021. CLC 1618, Genbank MT583321.

Russula intermedia P. Karst.

Cap 3–6.5 cm wide, convex, shallow convex, deep red, reddish brown, orange-brown, mottled with pink, ocher, cream, and yellow tones, smooth, greasy, or sticky; margin not striate, turned down; cuticle thick, hardly peeling. **Gills** narrowly attached, adnate, crowded, white, cream, then golden yellow. **Stem** 2–5 × 1–2 cm at apex, 1–2.5 cm at base, clavate, white, smooth, matte. **Flesh** white and solid. **Odor** absent, or fruity. **Taste** mild in our samples, but usually acrid. **Spores** yellow, 7.5–8.5 × 6.5–7.5 µm, subglobose to broadly ellipsoid, with isolated amyloid warts.

Ecology and Distribution In the transition zone between subalpine and alpine habitats in Alaska with birch, *Dryas,* and willow, although it is primarily known from montane to subalpine habitats with birch in Northern Europe, especially Finland. Possibly in aspen-cottonwood forests in Montana and the Pacific Northwest of North America. August.

Notes Easily recognized by the orange cap and hot taste (when present), but the cap can also be deep red, in which case it can be confused with other red Russulas. Also, the taste in ours was mild. *Russula aurantiolutea* Kauffman may be the same and has been reported from eastern and western United States and Canada.

- Cap orange, mottled with other colors (some deep red)
- Gills cream to dark ocher; stipe white
- Taste very hot, occasionally mild
- Spores yellow, with amyloid warts
- With birch or *Dryas*, possibly also with willows, aspen

Reference Noffsinger 2020. CLC 3822, Genbank MT500711.

Russula laccata Huijsman

Cap 1.5–4.5 cm across, convex to almost flat, sometimes depressed in center, deep magenta, deep maroon, with blackish center, smooth, greasy; margin not striate. **Gills** narrowly attached, close to distant, white, cream. **Stem** 1.5–5.0 × 0.5–1.5 cm, clavate, white, fragile. **Flesh** white. **Odor** absent, or faintly fragrant. **Taste** slightly to strongly acrid. **Spores** white, 7.5–8.5 × 5.5–6.5 µm, ellipsoid, with amyloid ridges.

Ecology and Distribution In North America, reported from Colorado, Wyoming, and Montana. Primarily with dwarf willow *Salix reticulata* and shrub willow *S. planifolia,* in August. Widely distributed in the Arctic and high alpine in the Northern Hemisphere with dwarf and shrub willows.

Notes *Russula laccata* is common in wet willow areas of the Rocky Mountains; it has a maroon cap with an almost black center and tastes hot. Of the other two dark maroon alpine species, *Russula saliceticola* lacks a hot taste and often has a pink blush on the stem; *Russula purpureofusca* has dark yellow gills and spores. *Russula laccata* is occasional in the subalpine with willows.

- Cap deep maroon with black center
- Gills and stem white
- Taste hot; odor mild
- Spores white, with amyloid ridges
- With willows

Reference Noffsinger and Cripps 2021. CRN 150, Genbank MT583211.

Russula laevis Kälviäinen, Ruotsalainen & Taipale

Cap 4–8 cm wide, concave, depressed in center, white, dingy white, cream, mottled with orange-brown stains, smooth, greasy, kidskin, or finely tomentose (use hand lens), with debris adhering; margin turned under. **Gills** narrowly adnate, subdecurrent, close, fragile, white to pale yellowish white, occasionally with greenish or grayish cast. **Stem** 1–3 × 1–2 cm, equal, short, white, staining ocher, matte, minutely pubescent at apex (use hand lens), smoother below. **Flesh** white, firm, solid, compact. **Odor** fruity or sweet, unpleasant in old specimens. **Taste** acrid or slowly acrid. **Spores** cream, 8–9.4 × 6.5–8 µm, subglobose with isolated, amyloid warts.

Ecology and Distribution In Arctic and alpine areas, often slightly buried in willow mats or *Dryas* debris. Reported from Colorado, and confirmed from Finland, likely widespread in Arctic-alpine habitats but previously reported as *R. delica.* August.

Notes This small, white *Russula* is reminiscent of a *Lactarius* with its sunken center but does not have latex. It looks like a miniature *Russula brevipes,* which is a hard-fleshed subalpine species that is usually much larger, although we did find a small form of *Russula brevipes* at treeline in Alaska (on Eagle Summit).

- Cap white, sunken in center
- Gills whitish, stem whitish and short
- Flesh hard; taste hot; odor fruity
- Spores cream, with amyloid warts
- Often buried in *Dryas* duff, or with willows

Reference Noffsinger and Cripps 2021. CLC 1642, Genbank MT500695.

Russula nana Killerm.

Cap 1.5–3.5 cm across, broadly convex to almost flat, candy apple red, paling to white from center outward, sometimes with ocher tints, smooth, slightly greasy; margin short striate or not. **Gills** narrowly attached, close, white. **Stem** 1.5–2.5 × 0.5–1.5 cm, clavate, white, smooth. **Flesh** white. **Odor** not distinct. **Taste** slightly acrid. **Spores** white, spores 7–8 × 6–6.5 µm, broadly ellipsoid, with amyloid ridges.

Ecology and Distribution In North America, reported from Colorado, Wyoming, and Montana, primarily with dwarf willow *Salix reticulata*. Widely distributed in cold climates such as the Arctic and mountaintops above treeline with dwarf willows. August.

Notes The small, red cap fading to white, white spore print, and association with dwarf willows help identify this species. *Russula laccata* and *R. saliceticola* have darker maroon caps with a blackish center. The similar but larger *Russula montana* is closely related and occurs with conifers in the krummholz zone.

- Cap cherry red, fading to white
- Gills and stem whitish
- Taste slightly hot; odor not distinct
- Spores with amyloid ridges
- With willows

Reference Noffsinger and Cripps 2021. CLC 3619, Genbank MT583306.

Russula neopascua Noffsinger & C.L. Cripps

Cap small, 2.5–3.5 cm across, irregular convex, becoming plane, deep magenta, dark red, with hints of green, viscid to dry, cuticle separable; margin striate for a short distance. **Gills** adnate-marginate, lamellulae absent, light straw yellow, yellow-brown. **Stem** 1.5–2 × 0.8–2 cm, clavate, white to off white, drying slightly pink. **Flesh** white. **Odor** fishy. **Taste** weakly acrid. **Spores** white to cream, 8–10 × 6.5–7.5 μm, with amyloid warts and spines.

Ecology and Distribution In the alpine with dwarf willow *Salix reticulata*. Only known from Niwot Ridge, Colorado, and the Beartooth Plateau, Montana. Currently thought to be endemic. August.

Notes The small size, magenta cap, yellowish gills, and fishy odor are the primary features for identification of *Russula neopascua*. *Russula pascua* is an alpine species from the Alps, which is related, and to *Russula nuoljae*. *Russula subrubens,* also with a fishy odor, is much larger and the cap is mottled orange-brown.

- Cap small, magenta, dark red, some with hints of green
- Gills yellowish; stem white or with pink tints
- Odor fishy; taste mildly acrid
- With dwarf willow *Salix reticulata*

Reference Noffsinger et al. 2024. CRN 146, Genbank MT583274.

Russula purpureofusca Kühner

Cap 2–4.5 cm across, convex becoming flat with slight depression, deep red, reddish brown, deep magenta, light yellowish green, yellow-brown, darker over disk in age, smooth, viscid, with debris adhering; margin not striate; cuticle separable except at center. **Gills** attached, adnate, cream, light yellow, dark yellow-cream, brownish yellow, elastic not brittle. **Stem** 1–3.5 × 0.5–1.5 cm, clavate, white, light yellow, graying, yellow-brown stains near base, smooth. **Flesh** hard or soft, white or watery gray. **Odor** absent, fungoid, or sweet. **Taste** strongly latently acrid, sometimes mild. **Spores** dark yellow, 8.0–9.0 × 6.5–8 μm, subglobose to broadly ellipsoid, with isolated, amyloid warts.

Ecology and Distribution In the alpine zone with shrub willow *Salix planifolia* and other willow species and *Dryas*; in North America occurring at least in Alaska and Colorado. Originally known from Finland. August.

Notes The deep magenta cap, yellow gills, and spores are diagnostic. *Russula laccata* has white gills and spores, and *Russula saliceticola* has pink on the stem and a mild taste. *Russula cupreola* may be the same species. As with other Russulas, cap color varies dramatically from yellow-brown (immature) to deep red, making the species more difficult to identify.

- Cap deep red to yellow-brown (immature); stem white
- Gills cream, yellow
- Taste usually acrid
- Spores dark yellow, with amyloid warts
- With willows

Reference Noffsinger and Cripps. 2021. CLC 3820, Genbank MT583254.

Russula saliceticola (Singer) Kühner ex Knudsen & Borgen

Cap 1.0–5.5 cm across, convex, occasionally depressed in center, deep red, ruby red, almost black in the center, smooth, greasy to dry, matte or shiny, often with debris adhering; margin slightly striate in age. **Gills** narrowly attached, white, cream, light yellow, sometimes colored reddish near cap margin. **Stem** 1.5–6.0 × 0.5–2.5 cm, slightly clavate, white, with or without a faint pinkish tint, which is more pronounced on drying. **Flesh** white, pink under cuticle. **Odor** absent, or faintly sweet. **Taste** mild. **Spores** white to cream, spores 8.5–9.5 × 6–7 μm, elliptical, with amyloid warts and ridges.

Ecology and Distribution In the Rocky Mountains, on the Beartooth Plateau in Montana and Wyoming; in Colorado on Niwot Ridge, and near Blue Lake, usually with the shrub willow *Salix planifolia*. Reported from Arctic and alpine areas of Finland, Norway, Sweden, and Iceland with willows. August.

Notes Often confused with *Russula laccata,* which has a deep magenta cap and black center, but white spores and a hot taste. *Russula saliceticola* is mild, and the stem often has a pink tint. *Russula purpureofusca* has a deep red cap, but yellow spores with individual warts.

- Cap deep ruby red with black center
- Gills light yellow; stem white, often with pink tint
- Taste mild
- Spores white to cream, with amyloid warts and ridges
- Primarily with shrub willow *S. planifolia*

Reference Noffsinger and Cripps 2021. CRN 173, Genbank MT583284.

Russula cf. *sphagnophila* Kauffman

Cap 2–3.5 cm wide, dished in center with uplifted margin, reddish maroon, rosy, with blackish center, lighter toward margin, viscid when wet, striate almost to center, evenly pleated; margin a bit crenate. **Gills** narrowly attached, a bit separated, cream, tinged pinkish in age. **Stem** 3.5–5 × 0.5 cm at apex to 1.3 cm at base, clavate, white, smooth with rosy-lavender tinge at apex. **Flesh** fragile. **Odor** absent. **Taste** not recorded. **Spores** cream, 8–9.5 × 6.5–7.5 µm, broadly ellipsoid, with prominent, amyloid warts, 1 µm high.

Ecology and Distribution In North America, at treeline with bog birch, near hemlock and spruce in Alaska. Also in the Altai Mountains of Russia. August.

Notes A fragile species with a rosy purple striated, dished cap and lavender-tinted stem found with birch. Originally described from Michigan (Kauffman in 1909) and subsequently reported by Singer from the Russian Altai Mountains; our ITS sequence matches Singer's collection. There is some controversy as to the taxonomy of this species.

- Cap thin, striate-pleated, rosy with black center, dished
- Stem fragile with lavender apex
- Gills white tinged pink
- Spores cream, with amyloid warts
- With birch

Reference Noffsinger 2020. CLC 3779, Genbank MT583279.

Russula subrubens (J.E. Lange) Bon

Cap 1.0–8.0 cm across, convex to almost flat, mottled with reddish orange, reddish brown, orange-brown, and yellow-brown colors, smooth, dry to greasy; margin not striate. **Gills** narrowly attached, crowded, becoming dark cream to yellowish, staining brown. **Stem** 1.0–6.0 × 1.0–3.0 cm, clavate, white to yellowish, smooth, dull, greasy. **Flesh** white, spongy to hard. **Odor** fishy, especially in older specimens. **Taste** mild to bitter. **Spores** yellow, 8–9 × 6–7 µm, ellipsoid, with amyloid warts.

Ecology and Distribution In North America, in alpine areas of Montana, Wyoming, and Colorado. Typically with shrub willow *S. planifolia* but often with dwarf willows nearby. Reported from alpine, subalpine, and Arctic habitats in Europe and Greenland. August.

Notes All our reports are from the alpine, but *R. subrubens* has been found in subalpine habitats in Europe with willows, which are often overlooked in conifer forests. The fishy odor and mottled cap colors help identify this species. It is one of the largest Russulas in the alpine. In the krummholz zone, *Russula xerampelina* also has a fishy odor but is more reddish.

- Cap mottled orange-brown colors
- Gills and spores yellow; stem white, robust
- Odor fishy
- Spores yellow, with amyloid warts
- Usually with shrub willows

Reference Noffsinger and Cripps 2021. CRN 137, Genbank MT583239.

5

A Few Small Non-Gilled Fungi in the Alpine

Numerous small non-gilled fungi are found in Arctic and alpine habitats, but these are not the focus of this book. Many are microscopic ascomycetes that do the work of decomposing plant material in cold climates; a whole body of literature is dedicated to these small fungi. *Dryas* hosts a whole cadre of microfungi inside its wood and leaves, as do willow, birch, and alder. Microfungi also can be found on and in dead grasses, and in stems and leaves of herbaceous plants in both wet and dry areas. Many plant pathogens, including the rusts and smuts, have adapted to the extreme climatic conditions. Species of *Typhula* (a tiny club-shaped fungus parasitic on grasses) have been shown to produce their own anti-freeze to withstand the cold conditions. There are even slime molds that inhabit these cold environments, although they are not common.

Bird's nest fungus *Crucibulum laeve* on willow stems.

The roots of woody alpine plants are colonized by the ectomycorrhizal fungi presented in this book. The roots of grasses and herbaceous plants host microscopic arbuscular mycorrhizal (AM) fungi, which lack fruiting structures and produce spores underground. Another group, the so-called dark septate fungi, infiltrate the roots of some Arctic-alpine plants and benefit them. The ericaceous plants, heathers and heaths, host their own set of mycorrhizal fungi (Cripps and Eddington 2005). Here we present a few small non-gilled Ascomycota and Basidiomycota that produce small fruiting structures and can be quickly identified. **A key** is provided in the introduction (pg 17). Fleshy, non-gilled puffballs and morels are treated in the section on fungi in meadows, and pored boletes are in the mycorrhizal section.

Bryoglossum gracile (P. Karst.) Redhead

Fruiting body 1–1.5 cm tall, club-shaped, matchstick-shaped. **Head** 2–3 mm across by 3–4 mm high, round to ovoid, pale yellow, yellow, pale orange-yellow, smooth, waxy. **Stem** 0.8–1.2 × 0.1 cm, equal, thin, undulating, white, pruinose; inserted into head. Difficult to tell if the edge of head is free. **Spores** 10–13 × 2–3 µm, blunt fusiform, some curved, sometimes septate.

Ecology and Distribution In mosses, often dead mosses in the Rocky Mountain alpine, and other Arctic-alpine habitats, inlcuding the Alps, Greenland, Fennoscandia, Iceland, Svalbard, Canada, and Alaska. Likely circumpolar. August.

Notes Often considered conspecific with the darker orange and more wrinkled *B. rehmii*; we elect to keep the two species separate pending further investigation. Some suggest that a free cap edge indicates this species, but we have found this difficult to determine.

- Fruiting body, small, shaped like a matchstick
- Head pale yellow, yellow, smooth
- Stem thin, white
- Inserted in mosses, often dead mosses

References Jamoni 2008; Schumacher and Jenssen 1992. CLC 3591.

Bryoglossum rehmii (Bres.) Ohenoja

Fruiting body 0.5–2 cm tall. **Head** 0.3–0.6 cm tall by 0.2–0.3 cm wide, irregularly club-shaped or somewhat spathulate, wrinkled, mottled light and dark orange; edge not connected to stem. **Stem** 0.6–1.0 × 0.1 cm, equal, pale yellow to pale orange, smooth, perhaps pruinose where stem meets the cap; base of stem inserted in moss. **Spores** 9–14 × 2–3 μm, fusiform, light-colored, some septate.

Ecology and Distribution Common on the Beartooth Plateau in moss parts that may be alive or dead. Together with *B. gracile*, there is likely a circumpolar distribution with records from Fennoscandia, Iceland, Greenland, Svalbard, Canada, and the Alps. August.

Notes *Bryoglossum gracile* is similar and is usually described as more yellow (less orange), more ovoid (not wrinkled), with a cap margin that does not touch the stem. Our collection better fits *B. rehmii,* although the cap margin is free on some.

- Fruiting body tiny, club or matchstick-shaped
- Head a mottled orange color, wrinkled
- Stem pale yellow
- Inserted deeply in mosses

References Schumacher and Jenssen 1992; Jamoni 2008. CLC 3898.

Ciborinia **cf.** *ciborium* (Vahl) T. Schumach. & L.M. Kohn

Cup 0.5–1.0 cm, deeply cup-shaped, goblet-shaped, tapering underneath. **Interior and exterior** brown, rust brown, dark ocher, smooth; rim darker or not. **Stem** 1–3 × 0.1–0.2 cm, yellow-brown, but black at the base, smooth. **Sclerotium** a flattened obscure disc shape. **Spores** 14–16 × 5–6 µm, light-colored, ellipsoid with tapered ends, smooth.

Ecology and Distribution Reported from the Montana alpine, fruiting on mosses in wet seeps under willows. An Arctic-alpine species with a circumpolar distribution, more commonly reported from the high Arctic (Canada, Greenland, Svalbard, Ellsemere Island) on *Eriophorum* (cottongrass) and *Carex* (sedges).

Notes A sclerotium was not noticed for this collection but could have been buried or overlooked. *Eriophorum* was not present, but *Carex* was nearby; ours appeared attached to moss. Synonymous with *Sclerotinia arctica*.

- Cups brown, rust brown, dark ocher, smooth
- Stem yellow-brown, black at the base, smooth
- Should have sclerotia
- Spores 14–16 × 5–6 µm
- In mosses, in wet sedge and willow areas

References Schumacher and Jenssen 1992; Schumacher and Kohn 1985. CLC 1795.

Scutellinia* cf. *minor (Velen.) Svreček

Fruiting body 2–5 mm across, disc-shaped. **Interior** reddish orange, smooth; cup margin fringed with brown hairs sticking out. **Exterior** reddish orange, covered with brown hairs. **Spores** hyaline, round to almost round, 10–15 μm, thick-walled, appearing smooth when immature; should be 15–20 μm and warted when mature. **Asci** up to 200 μm long, not amyloid. **Hairs** up to 500 × 10–14 μm, pointed, thick-walled, septate, brown.

Ecology and Distribution Here reported from the San Juan Mountains in Colorado at 3,690 m. Also known from the low alpine zone in the Alps, Iceland, Scotland, Switzerland, and Czechoslovakia. July, August.

Notes Colorado specimens were immature, so the species cannot be absolutely confirmed. However, only a few "eyelash cups" that have round spores have been consistently found in AA habitats. Another is *S. hyperborea,* which has more densely warted spores and cannot be ruled out.

- Fruiting body a small orange-red disc
- Margin of disc with straight brown hairs
- Spores almost round and warted when mature
- Ours was on soil in wet moss

References Jamoni 2008; Schumacher and Jenssen 1992. CLC 1878.

Basidiomycota

Clavaria argillacea Pers.

Fruiting body club-shaped with rounded top, finger-like, sometimes flattened, 1–5 cm high × 0.2–0.6 cm wide, dull yellow, pale yellow, smooth; often clustered. **Spores** white, 9–12 × 5–6 µm, long ellipsoid to cylindrical; with clamped hyphae.

Ecology and Distribution Reported from subarctic habitats in southern Canada near Churchill and Alaska in North America. Also reported from cold, acid heaths in Scotland, Greenland, Russia, and lower elevations in Scandinavia and Europe. August.

Notes One of the few coral fungi that occurs in heaths; it may have a relationship with ericaceous plants. It is recognized by its yellow color, habitat, and long spores. Common and taller at lower elevations in Europe.

- Club or finger-shaped, in clusters
- Pale to dull yellow
- Long ellipsoid spores
- In ericaceous heaths

Reference Ohenoja and Ohenoja 2010. Photo from Greenland.

Crucibulum cf. *laeve* (Huds.) Kambly

Fruiting body tiny, 1–2 mm wide, vase-shaped, woody-papery. **Lid** cushion-shaped, bright golden orange, textured. **Exterior** dingy brown, yellow-brown, red-brown, with hairs. **Interior** grayish cream, smooth. **Peridoles** (eggs) 0.5 mm, white, gray, oval, smooth, attached with a cord in a gel; 4–8 per fruiting body. **Spores** 8–10 × 4 μm, ovoid, smooth.

Ecology and Distribution Found on shrub willow branches in the alpine from Montana to Colorado. Also reported from Arctic Greenland on willow. August.

Notes At first thought to be rare, but numerous collections suggest that this species occurs regularly in alpine situations. However, fruiting bodies were extremely tiny, 1–2 mm across, which fits the size of *C. parvulum*. But the color of the lid and size of peridioles and spores match that of *C. laeve*; *C. crucibuliforme* is a synonym and may be the correct name.

- Fruiting body vase-shaped
- Lid golden orange
- Peridoles (eggs) white to gray
- On willow branches

References Borgen 1993; Borgen et al. 2006. CLC 3898.

Melampsora epitea Thüm. s.l.

Description This rust (Uredinales) appears as orange pustules of urediniospores on dwarf willow *Salix reticulata*. No collections were made, precluding direct observation of spores. However, urediniospores are typically round to ellipsoid, 14–24 × 11–20 µm, and spiny.

Ecology and Distribution Occasional on leaves of *Salix reticulata* in the Wyoming and Montana alpine zone. *Melampsora epitea* is a rust that is widely distributed in the Northern Hemisphere with willows. It has also been reported on dwarf willows, including in the Alps, Greenland, Canada, and Alaska.

Notes This rust is now known to be a complex of species that also includes *Melamopsora alpina, M. arctica,* and *M. reticulatae* as synonyms or separate species.

- Powdery orange pustules on leaves of dwarf willows

References Borgen et al. 2006; Zhao and Cai 2017; Durrieu and Adhikari 1993; Bennett et al. 2011.

Glossary

acrid—hot taste

adnate—broadly attached (as in gills)

adnexed—narrowly attached (as in gills)

ampul-shaped—narrow on top, swollen on bottom

amyloid—blue in Melzer's iodine solution

anastomosing—veinlike, fusing (as in gills)

annulus—ring (on the stem)

appressed (scales)—flattened scales

areolate—with a cracking pattern

ascus, asci—sac-like cell containing ascospores

attached gills—touching the stem

basidium, basidia—cell(s) with basidiospores on top

bulbous—swollen (stem base)

calyptra—loose sac (surrounding a spore)

campanulate—bell-shaped (cap)

capillitium—thick-walled hyphae (in a puffball)

capitate—with swollen head-like top (cystidia)

caulocystidia—sterile cells on stem

cespitose—clustered (as in fruiting bodies)

cheilocystidia—sterile cells on gill edges

clavate—club-shaped (as in stem or cystidia)

concolorous—same color (gill edge or cap/stem)

conic-convex—cap pointed and rounded

concave—sunken in center (cap)

convex—rounded out (like a mushroom cap)

cortina—cobwebby veil from cap edge to stem

crenate—scalloped (cap margin)

chrysocystidia—cystidia with yellow contents in KOH

cuticle—skin of a mushroom cap

cystidia—sterile cells (on gill, cap, or stem)

decurrent—running down the stem (as in gills)

dextrinoid—turning red in Melzer's iodine solution

domed—with a raised center (cap)

emarginate—gills notched at stem

endoperidium—inner layer (puffball)

exoperidium—outer layer (puffball)

farinaceous—mealy odor

fibrillose—with obvious fibers (cap/stem)

fibrous—with less obvious fibers

fimbriate—fibrous, fringed (as in cap)

floccules—with flakes (on a stem)

free gills—not touching the stem

fusoid—spindle-shaped, tapering (as in cystidia)

germ pore—opening on the end of a spore

gleba—interior spore-producing area (puffball)

globose—round

glutinous—slimy

granular—with bran-like particles (on cap)

hoary—with whitish coating (cap/stem)

hygrophanous—drying a lighter color

isodiametric—same size both ways (as in spores)

ITS region—section of DNA that defines species

krummholz—treeline area of low deformed trees

lageniform—flask shaped (cystidia)

latex—a milky substance (in gills of *Lactarius*)

marginate—rimmed (stem base)

membranous—tissue-like (as in a ring)

mycelium—the threadlike body of a fungus

necropigment—dark granular pigment (in basidia of *Mallocybe*)

nodulose—with distinct lumps (on spores)

ostiole—a pore (in puffballs)

papilla—a small bump (on top of a cap)

pedicel—piece of sterigma attached to a spore; or the narrow base of cystidia

peridiole—a container of spores

peridium—outer layer (in puffballs)

peronate—sock-like sheath (stem)

phaeseoliform—bean-shaped (spore)

pileipellis—skin of the cap (microscopically)

pileocystidia—sterile cells sticking up on cap

pleurocystidia—cystidia on the sides of gills

pruinose—surface with short hairs or cystidia (cap/stem)

pubescent—velvety

raphanoid—radish-like (odor)

sclerotium, sclerotia—hard knot of mycelium; resting stage of a fungus

septate—divided by a cell wall

spores—reproductive propagules of fungi

sterigma—elongation on a basidium

sterile base—lower area without spores (puffball)

striate—with distinct lines (cap/stem)

subcapitate—almost with a head (cystidia)

subglobose—almost round

superior—on the top part (of stem)

tibiiform—with round top, narrow neck, and swollen base

tomentose—rough wooly (cap/ stem)

umbo—a bump or raised area (on cap center)

urediniospores—red spores of a rust fungus

utriform—bag-shaped (cystidia)

veil—tissue covering parts of a fruiting body

velipellis—tissue overlaying cap surface

verrucose—with a warty surface (spores)

vinaceous—reddish wine-colored

viscid—slimy

References

Antonín V, Ďuriška O, Gafforov Y, Jančovičová S, Para R, Tomšovský M. 2017. Molecular phylogenetics and taxonomy in *Melanoleuca exscissa* group, (Tricholomataceae, Basidiomycota) and the description of *M. griseobrunnea* sp. nov. Plant Systematics and Evolution 303: 1181–1198.

Armada F, Bellanger J-M, Moreau PA. 2024. Champignons de la zone alpine. Contribution à l'étude des champignons supérieurs alpins, Annemasse, FMBDS. Pp. 376.

Barge E, Cripps CL. 2016a. Systematics of the ectomycorrhizal genus *Lactarius* in the Rocky Mountain alpine zone. Mycologia 108(2): 414–440.

Barge E, Cripps C. 2016b. New reports, phylogenetic analysis, and a key to *Lactarius* Pers. In the Greater Yellowstone Ecosystem informed by molecular analysis. MycoKeys 15:1–58.

Beker H, Eberhardt U, Vesterholt J. 2010. *Hebeloma hiemale* Bres. In Arctic/Alpine habitats. North American Fungi 5: 51–65.

Beker H, Eberhardt U, Vesterholt J. 2016. *Hebeloma*. Fungi Europaei 14. Edizioni Tecnografica, Lomazzo, Italia. Pp. 1217.

Bennett C, Aime MC, Newcombe G. 2011. Molecular and pathogenic variation within *Melampsora* on *Salix* in western North America reveals numerous cryptic species. Mycologia 103(5): 1004–1018.

Boertmann D. 1995. The genus *Hygrocybe*. Vol. 3. The Danish Mycological Society, Denmark. Pp. 184.

Bon M. 1985. Quelques taxons nouveaux pour la flore mycologique alpine. Bull. trimest. Féd. Mycol. Dauphiné-Savoie 25(97): 23–30.

Bon M. 1988. Quelques Agaricomycetes intéressants de la zone alpine récoltés dans le Tessin. Mycologia Helvetica 3(3): 315–330.

Bon M. 1988. Notes sur quelques récoltes intéressantes faites au stage de Pralognan et alentours (25–31 Août 1987). Bull. Féd. Mycol. Daupiné-Savoie-Juillet 110: 13–15.

Bon M. 1992. Quelques Inocybes alpins au stage de mycologie des arcs. Bull. trimest. Féd. Mycol. Dauphiné-Savoie 126: 19–22.

Borgen T. 1993. Svampe I Grønland. Atuakkiorfik, Nuuk (Greenland). Pp. 112.

Borgen T, Elborne S, Knudsen H. 2006. A checklist of the Greenland basidiomycetes. Meddelelser om Grøenland Bioscience 56: 37–59.

Brandrud TE, Bendiksen E, Jordal JB, Weholt Ø, Eidissen SE, Lorås J, Dima B, Noordeloos ME. 2018. *Entoloma* species of the rhodopolioid clade (subgenus *Entoloma*; Tricholomatinae, Basidiomycota) in Norway. Agarica 38: 21–46.

Breitenbach J, Kränzlin F. 2000. Fungi of Switzerland Vol 5: Cortinariaceae. Verlag Mykologia, Luzern, Switzerland. Pp. 338.

Bunyard B, Justice J. 2020. Amanitas of North America. The FUNGI Press, Batavia, Illinois. Pp. 336.

Buyck B et al. 2022. Fungal biodiversity profiles 111–120. Cryptogamie Mycologie 43(2): 23–61. (#112)

Clowez P, Moreau PA. 2020. Morilles de France et d'Europe. Cap Régions Éditions, Noyon, France. Pp. 370.

Coker WC, Couch JN. 1974 (1928). The Gasteromycetes of the Eastern United States and Canada. Dover Publications, New York.

Cripps CL, Eddington LH. 2005. Distribution of mycorrhizal types among alpine vascular plant families on the Beartooth Plateau, Rocky Mountains, USA, in reference to large-scale patterns in arctic-alpine habitats. Arctic, Antarctic, and Alpine Research 37(2): 177–188.

Cripps CL, Horak E. 2006. *Arrhenia auriscalpium* in arctic-alpine habitats: world distribution, ecology, new reports from the southern Rocky Mountains, USA. Meddelelser om Grøenland Bioscience 56: 17–24.

Cripps CL, Horak E. 2010. *Amanita* in the Rocky Mountain alpine zone, USA: New records for *A. nivalis* and *A. groenlandica*. North American Fungi 5: 9–21.

Cripps CL, Larsson E, Vauras J. 2020. Nodulose-spored *Inocybe* from the Rocky Mountain alpine zone molecularly linked to European and type specimens. Mycologia 112: 133–153.

Cripps CL, Eberhardt U, Schütz N, Beker HJ, Evenson VS, Horak E. 2019. The genus *Hebeloma* in the Rocky Mountain Alpine Zone. MycoKeys 46: 1–54.

Davey ML, Heimdal R, Ohlson M, Kauserud H. 2013. Host- and tissue-specificity of moss-associated *Galerina* and *Mycena* determined from amplicon pyrosequencing data. Fungal Ecology 6: 179–186.

Den Bakker H, Zuccarella G, Kuyper T, Noordeloos M. 2007. Phylogeographic patterns in *Leccinum* sect. *Scabra* and the status of the arctic/alpine species *L. rotundifoliae*. Mycological Research 111: 663–672.

Desjardin DE, Wood M, Stevens FA. 2015. California Mushrooms. Timber Press, Inc. Portland, OR. Pp. 559.

Dima B, Liimatainen K, Niskanen T, Bojantchev D, Harrower E, Papp V, Nagy LG, Kovács GM, Ammirati JF. 2021. Type studies and fourteen new North American species of *Cortinarius* section *Anomali* reveal high continental species diversity. Mycological Progress 20: 1399–1439.

Durrieu G, Adhikari M. 1993. Distribution of some groups of plant parasitic fungi in high mountains. Bibl. Mycol. 150: 23–31. [Arctic and Alpine Mycology 3]

Eberhardt U, Schütz N, Bartlett P, Beker HJ. 2022. 96 North American taxa sorted—Peck's *Hebeloma* revisited. Mycologia 114: 337–387.

Eberhardt U, Schütz N, Bartlett P, Beker HJ. 2023. Many were named, but few are current: The *Hebeloma* of Hesler, Smith and coauthors. Mycologia 115: 813–870.

Esteve-Raventós F, Gonzalez V, Arenal F, Horak E. 1998. Taxonomical remarks to *Collybia hariolorum* var. *alpicola* (Tricholomataceae) recently reported from the alpine zone of the Spanish Pyrenees and the French Alps. Zeitshrift für Mykologie Band 64/1: 67–72.

Favre J. 1948. Les associations fongiques des haute-marais jurassiens, Matériaux Fl. Cryptog. suisse 10 fasc 3 Berne.

Favre J. 1955. Les champignons supérieurs de la zone alpine du Parc National Suisse. Resultats des recherches scientifiques enterprises au Parc National Suisse. Ergebnisse der wissenschaftlichen Untersuchungen des schweizerischen National Parks 5: 1–212.

Ferrari E. 2006. Fungi non delineati I. *Inocybe* alpine e subalpine. Massimo Candusso, Italy. Pp. 457.

Gillman LS, Miller OK Jr. 1977. A study of the boreal, alpine and arctic species of *Melanoleuca*. Mycologia 69: 927–951.

Grund DW, Stuntz DE. 1968. Nova Scotian Inocybes. I. Mycologia 60: 407–425.

Grund DW, Stuntz DE. 1977. Nova Scotian Inocybes. IV. Mycologia 69: 392–408.

Gulden G. 2010. Galerinas in cold climates. North American Fungi 5: 127–157.

Gulden G, Jenssen KM. 1988. Arctic and Alpine Fungi—2. Soppkonsulenten, Oslo. Pp. 58.

Harder C, Læssøe T, Frøslev T, Ekelund F, Rosendahl S, Kjøller R. 2013. A three-gene phylogeny of the *Mycena pura* complex reveals 11 phylogenetic species and shows ITS to be unreliable for species identification. Fungal Biology 117(11–12): 764–775.

He ZM, Chen ZH, Bau T, et al. 2023. Systematic arrangement within the family Clitocybaceae (Tricholomatineae, Agaricales): phylogenetic and phylogenomic evidence, morphological data and muscarine-producing innovation. Fungal Diversity 123: 1–47.

Høiland K. 1984. *Cortinarius* subgenus *Dermocybe* with special regard to the species in the Nordic countries. Opera Botanica 71: 1–113.

Horak E. 1987. *Asterosporina* in the alpine zone of the Swiss National Park (SNP) and adjacent regions. Pp 205–234. In: Laursen GA, Ammirati JF, Redhead SA, eds. Arctic and Alpine Mycology 2, Plenum Press, NY.

Horak E, Miller OK Jr. 1992. *Phaeogalera* and *Galerina* in arctic-subarctic Alaska (U.S.A.) and the Yukon Territory (Canada). Can. J. Bot. 70: 414–433.

Hutchison L, Summerbell RC, Malloch D. 1988. Additions to the mycota of North America and Quebec: Arctic and Boreal species from Schefferville, North Quebec. Naturaliste Can. (Rev. Ecol. Syst.) 115: 39–56.

Jalink L. 2010. Additional notes on the Lycoperdaceae of the Beartooth Plateau. North American Fungi 5: 173–179.

Jamoni PG. 2008. Fungi alpini, delle zone alpine superiori e inferiori. Associazione Micologica Bresadola, Fondazione centro Studi Micologici, Trento. Pp. 543.

Jeppson M. 2018. Puffballs of Northern and Central Europe. Mykologiska Publikationer 8, Sweden. Pp. 360.

Kasuya T. 2010. Lycoperdaceae (Agaricales) on the Beartooth Plateau, Rocky Mountains, U.S.A. North American Fungi 5: 159–171.

Kerrigan R.W. 2016. *Agaricus* of North America. New York Botanical Garden Press, New York. Pp. 572.

Knudsen H, Borgen T. 1987. Agaricaceae, Amanitaceae, Boletaceae, Gomphidiaceae, Paxillaceae and Pluteaceae in Greenland. Pp. 235–253. In: Laursen GA, Ammirati JF, Redhead SA, eds. Arctic-alpine Mycology 2. Plenum Press, NY. Pp. 364.

Knudsen H, Vesterholt J, eds. 2008. Funga Nordica: Edition 1. Nordsvamp, Copenhagen. Pp. 965.

Kokkonen K. 2015. A survey of boreal *Entoloma* with emphasis on the subgenus *Rhodopolia*. Mycol Progress 14: 116.

Korotkin HB, Swenie RA, Miettinen O, Budke JM, Chen KH, Lutzoni F, et al. 2018. Stable isotope analyses reveal previously unknown trophic mode diversity in the Hymenochaetales. American Journal of Botany 105: 1869–1887.

Kühner R. 1978. Agaricales de la zone alpine. Genre *Melanoleuca* Pat. Bull. Mens. Soc. Linn. D. Lyon. 47: 12–52.

Kühner R. 1988. Diagnoses de quelques nouveaux Inocybes récoltes en zone alpine de la Vanoise. Doc. Mycol. 19(74): 1–27.

Kuyper TW. 1986. A revision of the genus *Inocybe* in Europe I. Subgenus *Inosperma* and the smooth-spored species of subgenus *Inocybe*. Persoonia Supplement 3: 1–247.

Lamoure D. 1972. Agaricales of the alpine zone, genus *Clitocybe*. Trav. Sci. Nat. Van. 2: 107–152.

Largent DL. 1994. Entolomatoid fungi of the Western United States and Alaska. Mad River Press, Eureka, CA. Pp. 495.

Larsson E, Jeppson M. 2008. Phylogenetic relationships among species and genera of *Lycoperdaceae* based on ITS and LSU sequence data from north European taxa. Mycological Research 112(1): 4–22.

Larsson E, Jeppson M, Larsson K-H. 2009. Taxonomy, ecology, and phylogenetic relationships of *Bovista pusilla* and *B. limosa* in North Europe. Mycological Progress 8(4): 289–299.

Larsson E, Vauras J, Cripps CL. 2014. *Inocybe leiocephala*, a species with an intercontinental distribution range—disentangling the *I. leiocephala-subbrunea-catalaunica* morphological species complex. Karstenia 54: 15–39.

Larsson E, Vauras J, Cripps CL. 2018. *Inocybe praetervisa* group—A clade of four closely related species with partly different geographical distribution ranges in Europe. Mycoscience 59: 277–287.

Larsson E, Vauras J. 2023. Fungal Planet description sheets: 1550–1613. Persoonia 51: 280–417. [No. 1575].

Liimatainen K, Cartere X, Dima B, Kytovuori I, Bidaud A, Reumaux P, Niskanen T, Ammirati JF, Bellanger J-M. 2017. *Cortinarius* section *Bicolores* and section *Saturnini* (Basidiomycota, Agaricales), a morphogenetic overview of European and North American species. Persoonia 39: 175–200.

Liimatainen K, Niskanen T, Ammirati JF, et al. 2015. *Cortinarius,* subgenus *Telamonia,* section *Disjungendi,* cryptic species in North America and Europe. Mycol. Progress 14: 1016.

Liimatainen K, Niskanen T, Dima B, Ammirati JF, Kirk P, Kytovuori I. 2020. Mission impossible completed: unlocking the nomenclature of the largest and most complicated subgenus of *Cortinarius*, *Telamonia*. Fungal Diversity 104: 291–331.

Matheny PB, Kudzma LV, Graddy MG, Mardini SM, Noffsinger CR, Swenie RA, Walker NC, Campagna SR, Halling R, Lebeuf R, Kuo M, Lewis DP, Smith ME, Tabassum M, Trudell SA, Vauras J. 2023. A phylogeny for North American *Mallocybe* (*Inocybaceae*) and taxonomic revision of eastern North American taxa. Fungal Systematics and Evolution 12: 153–201.

Matheny PB, Swenie RA. 2018. The *Inocybe geophylla* group in North America: a revision of the lilac species surrounding *I. lilacina*. Mycologia 110: 618–634.

Moser M, Horak E. 2006. *Agrocybe praemagna*: a new alpine species from Colorado, Idaho and Wyoming, Rocky Mountains, USA. Meddelelser om Grønland Bioscience 56: 133–138.

Moser M, McKnight KH. 1987. Fungi (Agaricales, Russulales) from the alpine zone of Yellowstone National Park and the Beartooth Mountains with special emphasis on *Cortinarius*. Pp. 299–317. In: Laursen GA, Ammirati JF, Redhead SA, eds. Arctic and Alpine Mycology 2. Plenum Press, New York, NY.

Moser M. 1993. Studies on North American Cortinarii. III. The *Cortinarius* flora of dwarf and shrubby *Salix* associations in the alpine zone of the Wind River Mountains, Wyoming, USA. Sydowia 45: 275–306.

Moser M, McKnight KH, Ammirati JF. 1995. Studies on North American Cortinarii I. New and interesting taxa from the greater Yellowstone area. Mycotaxon 60: 301–346.

Niskanen T. 2014. Nomenclatural novelties. Index Fungorum. 186: 2.

Niskanen T. 2014. Nomenclatural novelties. Index Fungorum. 197: 5.

Noffsinger C, Cripps CL, Horak E. 2020. A 200-year history of arctic and alpine fungi in North America: Early sailing expeditions to the molecular era. Arctic, Antarctic, and Alpine Research 52: 323–340.

Noffsinger C. 2020. Systematic analysis of *Russula* in the North American Rocky Mountain alpine zone. M.S. Thesis, Montana State University, Bozeman, MT. Available online.

Noffsinger C, Cripps CL. 2021. Systematic analysis of *Russula* in the North American Rocky Mountain alpine zone. Mycologia 113(6): 1278–1315.

Noffsinger C, Adamčíková K, Eberhardt U, Caboč M, Bazzicalupo A, Buyck B, Kaufmann H, Weholt Ø, Looney BP, Matheny PB, Berbee ML, Tausan D, Adamčík S. 2024. Three new species in *Russula* subsection *Xerampelinae* supported by genealogical and phenotypic coherence. Mycologia 116(2): 322–349.

Noordeloos ME. 1981. *Entoloma* subgenera *Entoloma* and *Allocybe* in the Netherlands and adjacent regions with a reconnaissance of its remaining taxa in Europe. Persoonia 11: 153–256.

Noordeloos ME et al. 2022. *Entoloma*. Flora Agaricina Neerlandica Vol. 1, Supplement. Candusso Editrice, Italy. Pp. 968.

Ohenoja E, Ohenoja M. 2010. Larger fungi of the Canadian Arctic. North American Fungi 5: 85–96.

Osmundson TW, Cripps CL, Mueller GM. 2005. Morphological and molecular systematics of Rocky Mountain alpine *Laccaria*. Mycologia 97: 949–972.

Peintner U. 1999. *Lepiota* and *Cystolepiota* (Agaricales) in Arctic-alpine habitats. Österr. Z. Pilzk. 8: 19–35.

Redhead S. 1984. *Arrhenia* and *Rimbachia,* expanded generic concepts, and a reevaluation of *Leptoglossum* with emphasis on muscicolous North American taxa. Canadian Journal of Botany 62: 865–892.

Ronikier A, Ronikier M. 2010. Biogeographical patterns of arctic-alpine fungi; distribution analysis of *Marasmius epidryas*, a typical circumpolar species in cold environments. North American Fungi 5: 23–50.

Saar I, Põldmaa K, Kõljalg U. 2009. The phylogeny and taxonomy of genera *Cystoderma* and *Cystodermella* (Agaricales) based on nuclear ITS and LSU sequences. Mycological Progress. 8(1): 59–73.

Sanchez-Garcia M, Cifuentes-Blanco J, Matheny PB. 2013. Taxonomic revision of the genus *Melanoleuca* in Mexico and description of new species. Rev. Mex. Biodivers. 84: S111–S127.

Schumacher T, Jenssen T-S. 1992. Arctic and Alpine Fungi—4. Soppkonsulenten, Oslo. Pp. 66.

Schumacher T, Kohn L. 1985. A monographic revision of the genus *Myrioslcerotinia*. Can. J. Bot. 63: 1610–1640.

Senn-Irlet B, Jenssen KM, Gulden G. 1990. Arctic and Alpine Fungi—3. Soppkonsulenten, Oslo. Pp. 58.

Siegel N, Schwarz C. 2016. Mushrooms of the Redwood Coast. Ten Speed Press, Berkeley. Pp. 602.

Smith AH. 1939. Certain species of *Inocybe* in the Herbarium of the University of Michigan. Pap. Mich. Acad. Sci. 24: 93–106.

Smith AH. 1947. North American species of *Mycena*. University of Michigan Press, Ann Arbor. Pp. 521.

Smith AH, Singer R. 1945. A monograph on the genus *Cystoderma*. Pap. Michigan Acad. Sci. 30: 71–124.

Vauras J, Larsson E. 2021. Fungal Planet description sheets 1182–1283. Persoonia 46: 313–528. [No. 1256].

Vesterholt J. 2005. The genus *Hebeloma*. Fungi of Northern Europe—vol. 3. Narayana Press, Gylling, Denmark. Pp. 146.

Vila J, Reschke K, Battisin E, Marulli U, Noordeloos ME, Moreau PA, Ribes M, Corriol G, Loizides M. 2020 (2021). New species of the genus *Entoloma* (Basidiomycota, Agaricales) from Southern Europe. Austrian J. of Mycol. 29: 123–153.

Voitk A, Saar I, Lücking R, Moreau PA, Corriol G, Thorn G, Hay C, Moncada B, Gulden G. 2020. Surprising morphological, ecological and ITS sequence diversity in the *Arrhenia acerosa* complex (Basidiomycota: Agaricales: Hygrophoraceae). *Sydowia* 73: 133–162.

Yu XD, Deng H, Yao YJ. 2011. *Leucocalocybe*, a new genus for *Tricholoma mongolicum* (Agaricales, Basidiomycota). African Journal of Microbiology Research 5: 5750–5756.

Zhao P, Kakishima M, Wang Q, Cai L. 2017. Resolving the *Melampsora* epitea complex. Mycologia 109: 391–407.

Photo Credits

All photographs by Cathy L. Cripps, with the exception of:

Ed Barge
Omphalina rivulicola
Lactarius pallidomarginatus

Henry Beker
Hebeloma alpinum
Hebeloma vaccinum

Vera Evenson
Amanita nivalis
Hebeloma subconcolor

Egon Horak
Cortinarius pratensis

Ellen Larsson/Jukka Vauras
Inocybe alpinomarginata
Inocybe murina
Mallocybe leucoblema
Cortinarius pulchripes

P. Brandon Matheny
Inocybe nemorosa

Chance Noffsinger
Russula laccata
Russula neopascua

Noah Siegel
Arrhenia auriscalpium
Hygrocybe conica
Lichenomphalia alpina
Mycena diosma
Rhizomarasmius epidryas
Russula intermedia
Russula sphagnophila

Taxonomic Index

Cathy L. Cripps is an emerita professor of mycology at Montana State University. She is the coauthor of *The Essential Guide to Rocky Mountain Mushrooms by Habitat* and the editor of *Fungi in Forest Ecosystems: Systematics, Diversity, and Ecology*.

The University of Illinois Press
is a founding member of the
Association of University Presses.

———————————————————

Text designed by Jim Proefrock
Composed in 8.75/12 Officina Serif
with Univers Condensed display
at the University of Illinois Press
Manufactured by Versa Press, Inc.

University of Illinois Press
1325 South Oak Street
Champaign, IL 61820-6903
www.press.uillinois.edu